SpringerBriefs in Applied Sciences and Technology

SpringerBriefs present concise summaries of cutting-edge research and practical applications across a wide spectrum of fields. Featuring compact volumes of 50 to 125 pages, the series covers a range of content from professional to academic.

Typical publications can be:

- A timely report of state-of-the art methods
- An introduction to or a manual for the application of mathematical or computer techniques
- A bridge between new research results, as published in journal articles
- A snapshot of a hot or emerging topic
- An in-depth case study
- A presentation of core concepts that students must understand in order to make independent contributions

SpringerBriefs are characterized by fast, global electronic dissemination, standard publishing contracts, standardized manuscript preparation and formatting guidelines, and expedited production schedules.

On the one hand, **SpringerBriefs in Applied Sciences and Technology** are devoted to the publication of fundamentals and applications within the different classical engineering disciplines as well as in interdisciplinary fields that recently emerged between these areas. On the other hand, as the boundary separating fundamental research and applied technology is more and more dissolving, this series is particularly open to trans-disciplinary topics between fundamental science and engineering.

Indexed by EI-Compendex, SCOPUS and Springerlink.

More information about this series at http://www.springer.com/series/8884

Amit Kumar Nayak · Md Saquib Hasnain

Plant Polysaccharides-Based Multiple-Unit Systems for Oral Drug Delivery

 Springer

Amit Kumar Nayak
Department of Pharmaceutics
Seemanta Institute of Pharmaceutical
Sciences
Mayurbhanj, Odisha, India

Md Saquib Hasnain
Department of Pharmacy
Shri Venkateshwara University
Amroha, Uttar Pradesh, India

ISSN 2191-530X   ISSN 2191-5318   (electronic)
SpringerBriefs in Applied Sciences and Technology
ISBN 978-981-10-6783-9   ISBN 978-981-10-6784-6   (eBook)
https://doi.org/10.1007/978-981-10-6784-6

Library of Congress Control Number: 2019934352

This Springer imprint is published by the registered company Springer Nature Singapore Pte Ltd.
The registered company address is: 152 Beach Road, #21-01/04 Gateway East, Singapore 189721, Singapore

About This Book

This volume presents almost all the overviews and recent topics related to various plant polysaccharide-based multiple-unit systems for oral drug delivery applications. The use of plant-derived materials in the drug delivery research is taken seriously into consideration today because of non-toxicity, biodegradability, readily availability, eco-friendly, and low extraction expenditure. Among various plant-derived materials, plant polysaccharides are a class of naturally occurring polymers present as storage carbohydrates in plants consisting of glucose monomers in cereals, root vegetables, rhizomes, seeds, fruits, etc. A substantial research endeavor has currently been directed to explore usefulness of various plant polysaccharides for pharmaceutical uses including drug delivery, and their use is being evolved from their auxiliary functions toward their active role as drug performance enhancers in terms of drug release, drug stability, bioavailability, target specificity, etc. During the past few decades, various plant polysaccharides have already been used to design oral multiple-unit sustained-release oral drug delivery systems like microparticles, beads, spheroids, etc. These multiple-unit systems have shown the capability to mix up with the gastrointestinal juices, to distribute over a wider area in the gastrointestinal tract, lowering both the intra- and inter-subject variability of the drug absorption with the reduction of dose dumping possibility.

Contents

About the Authors

Dr. Amit Kumar Nayak, Ph.D. is currently working as Associate Professor at Seemanta Institute of Pharmaceutical Sciences, Odisha, India. He has earned his Ph.D. in Pharmaceutical Sciences from IFTM University, Moradabad, U.P, India. He has over 10 years of research experience in the field of pharmaceutics, especially in the development and characterization of polymeric composites, polymeric gels, hydrogels, novel and nanostructured drug delivery systems. Till date, he has authored over 120 publications in various high impact peer-reviewed journals and 34 book chapters to his credit. Overall, he has earned highly impressive publishing and cited record in Google Scholar (H-Index: 32, i10–Index: 81). He has been the permanent reviewer of many international journals of high repute. He also has participated and presented his research work at several conferences in India and is a life member of Association of Pharmaceutical Teachers of India (APTI).

Dr. Md Saquib Hasnain, Ph.D. has over 6 years of research experience in the field of drug delivery and pharmaceutical formulation analyses, especially systematic development and characterization of diverse nanostructured drug delivery systems, controlled release drug delivery systems, bioenhanced drug delivery systems, nanomaterials and nanocomposites employing Quality by Design approaches as well as development and characterization of polymeric composites, formulation characterization and many more. Till date he has authored over 30 publications in various high impact peer-reviewed journals, 30 book chapters, and 1 Indian patent application to his credit. He is also serving as the reviewer of several prestigious journals. Overall, he has earned highly impressive publishing and cited record in Google Scholar (H-Index: 12). He has also participated and presented his research work at over ten conferences in India, and abroad. He is also the member of scientific societies i.e., Royal Society of Chemistry, Great Britain, International Association of Environmental and Analytical Chemistry, Switzerland and Swiss Chemical Society, Switzerland.

Abbreviations

ABR	Assam Bora rice
DSC	Differential scanning calorimetry
EDX	Electron dispersive X-ray
FM	Fenugreek seed mucilage
FTIR	Fourier transform infrared
GAr	Gum arabic
GIT	Gastrointestinal tract
GRAS	Generally recognized as safe
IPN	Interpenetrated polymer network
LG	Locust bean gum
LP	Linseed polysaccharide
OG	Okra gum
PS	Potato starch
SEM	Scanning electron microscopy
SG	Sterculia gum
TP	Tamarind polysaccharide
TS	Tapioca starch
XRD	X-ray diffraction

Chapter 1
Background: Multiple Units in Oral Drug Delivery

1.1 Introduction

The broad range of effectual drug candidates available in the present day is one of the best triumphs of the drug discovery research (Ansari et al. 2012; Hirani et al. 2009; Nayak and Manna 2011). As the new drug molecule development expenses are extremely high, the endeavors are currently being accomplished by the pharmaceutical industries to spotlight on the design and fabrication of various effective pharmaceutical dosage forms of existing drug molecules with desired efficacy as well as high level of safety mutually with the advantages of decreased dosing frequency and formulation expenditure (Jana et al. 2014; Nayak et al. 2010a, 2011a, 2013b, c, 2018b; Rath Adhikari et al. 2010). But, for the duration of past few decades, an augmented claim has been observed for the need of patient-friendly dosages (Das et al. 2017; Hasnain and Nayak 2018a; Jana et al. 2015a, b; Ray et al. 2018). Therefore, the drug delivery dosage systems are gradually fetching more sophistication as the drug delivery formulators and scientists attain much better understanding(s) of physical, chemical, physicochemical as well as biochemical factors, which are pertinent to the designing of effectual drug delivery dosage forms for different drug candidates to treat a variety of diseases (Hasnain and Nayak 2018a, b, 2019; Jana et al. 2014; Malakar et al. 2011, 2012a, b, c, d, 2014b; Nayak 2010; Nayak and Das 2018; Nayak and Jana 2013; Nayak and Sen 2009, 2016). Theoretically, a perfect drug delivery system must accomplish two most important prerequisite issues: (a) to facilitate the delivery of drug(s) at the preferred rate and (b) spatial drug targeting to the desired, specific, site(s) (Chein 1990; Nayak et al. 2018a). These important prerequisite issues elicit the need of controlled releasing of drug(s), which can facilitate the high-degree therapeutic effectiveness through the reduction of the number of required doses and dosage sizes (Pal and Nayak 2015a, b; Verma et al. 2004).

1.2 Oral Drug Delivery

In spite of the significant advancement in the research of drug delivery development and characterization, the oral route still remains the preferred and convenient most drug administration route because of administrative ease, low expenditure, and enhanced patient compliances (Malakar and Nayak 2013; Malakar et al. 2012d). From the viewpoints of the patients, the oral route of drug administration aids the painless self-medication procedure and is, therefore, regarded as the safest route of drug administration (Nayak et al. 2013a). To attain advantageous and desired pharmacokinetic results, it is very demanding to design and prepare effective oral drug delivery dosage forms optimizing drug stability in the gastrointestinal tract (GIT) (Verma et al. 2017). After the medication, drugs are immediately released from the conventional drug delivery dosage forms. However, the conventional drug deliveries of small molecular drugs are found not capable of providing the desired action of drugs over a prolonged period and the target specificity (Jana et al. 2013b; Nayak and Pal 2015). Oral administrations of conventional drug delivery dosage forms produce a wider range of fluctuations in the systemic drug concentrations associated with unfortunate effectiveness and the consequential undesirable side effects chances (Das and Das 2003; Malakar et al. 2014b, c). The continuance of drug concentrations in the systemic circulation of administered drugs within the therapeutic index is a very decisive issue for the successful treatment(s). This has also been revealed that the characteristics of drugs such as solubility, absorption, bioavailability, etc., are some essential factors for the effectual oral drug delivery outcome (Muller et al. 2001). Various carrier properties like residence time at a particular site are also considered as an important determinant for oral drug delivery. Various novel drug delivery dosage systems are being designed and developed to solve the above said shortcomings of the conventional drug delivery dosage forms through producing sustained releasing of drugs over a prolonged time, enhancing the drug solubility, maintaining the drug activity, minimizing side effect occurrences and facilitating the specificity for the drug target (Pawar et al. 2012). During the past few years, numerous research groups throughout the world are conducting research on designing, developing, and optimizing different novel drug delivery systems for the oral administration (Das and Das 2003).

Different orally administrable drug delivery systems can be categorized into two types: immediate releasing drug delivery systems and modified releasing drug delivery systems (Malakar and Nayak 2012b; Siepmann and Siepmann 2008). Immediate releasing drug delivery dosage forms are planned to disintegrate speedily and to produce immediate releasing of drug(s). These drug delivery systems are related to fluctuations in the plasma drug concentration levels. Sometimes, the fluctuations in plasma drug concentration levels unfortunately direct to the unsuccessful therapeutics with the side effect related risks (Chein 1990; Nayak et al. 2018a). The multiple dosing is necessary for immediate releasing drug delivery dosage forms to compensate for the decrease in the plasma drug concentration levels because of drug metabolisms and excretions. In contrast, various modified releasing drug

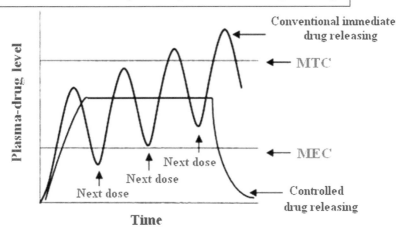

Fig. 1.1 Plasma-drug concentration profile for conventional immediate drug-releasing systems and controlled drug-releasing systems. MEC: minimum effective concentration and MSC: maximum safe concentration

delivery dosage forms have been designed for improving the pharmacokinetic profiles of different drug candidates and patient compliances with decreasing side effect occurrences (Getsios et al. 2004). The comparison of plasma drug concentration profile for conventional drug-releasing systems and controlled drug-releasing systems is presented in Fig. 1.1. The controlled drug-releasing presents some benefits over the conventional systems and these are as follows (Das and Das 2003; Nayak 2011; Vyas and Khar 2006):

(1) Avoidance of fluctuations of plasma drug concentration levels over a prolonged period to circumvent the sub-therapeutical and toxic concentrations for minimizing the unwanted side effects occurrences.
(2) Decreasing the administered doses while attaining the comparable effects.
(3) Reduction of dosing frequency.
(4) Targeting of drugs.

It is very much advantageous to design the modified drug delivery dosage forms, which release the drugs at programmed rates to accomplish the optimal drug concentrations at the desired sites of action (Vyas and Khar 2006). Modified drug delivery dosage forms for oral administration are generally employed for (Nayak 2011; Siepmann and Siepmann 2008):

(1) Delayed drug-releasing systems (e.g., by employing enteric coat);
(2) Extended, sustained and prolonged drug-releasing systems (e.g., drug releases controlling by first order, zero order, etc.);
(3) Programmed drug-releasing systems (e.g., pulsatile drug releasing, triggered drug releasing, etc.);

(4) Site-specific drug-releasing systems and timed drug-releasing systems (e.g., gastroretentive drug delivery, colonic drug delivery, etc.).

The delayed drug-releasing systems are differentiated from the other categories of modified drug delivery dosage forms as these are capable of exhibiting a well-defined lag time before the releasing of the drug(s) (Nayak et al. 2018a; Vyas and Khar 2006). In general, enteric-coated modified drug-releasing systems planned to travel through the stomach region are unaffected to produce drug releasing within the intestinal tract region. However, the extended, sustained, and prolonged releasing systems are the terms utilized almost synonymously for describing these groups of drug delivery dosage forms for controlled drug releasing with the desired reproducibility as well as predictability in the drug releasing kinetics (Longer and Robinson 1990). These controlled drug delivery dosage forms are also found capable to produce desirable systemic bioavailability with the reduction of systemic toxic side effects (Das and Das 2003; Pal et al. 2018).

The controlled drug releasing approach characterizes as one of the leading edge areas of science and technology concerning multidisciplinary advances for the healthcare applications (Das and Das 2003; Nayak 2011). Controlled drug-releasing systems can be described as some useful techniques or approaches, which are capable of delivering different drugs to the specified target sites at a period and rate designed to attain an intentional outcome (Das and Das 2003). These systems in drug delivery application areas not only lengthen the period of drug action, but also produce the therapeutic outcome related to the desired reproducible as well as a predictable drug-releasing kinetics. The purposes of the controlled drug-releasing systems comprise the supply of drug candidates through the specific areas of body devoid of any degradation and maintenance of therapeutically plasma drug concentration levels for an extended time with decreased dosing frequency, decreasing of side effects occurrences and improvement of patient compliances (Tiwari et al. 2009; Vyas and Khar 2006). Some of the important approaches to attain oral controlled drugs releasing drug delivery systems are as follows (Nayak 2011; Vyas and Khar 2006):

(1) Coating approaches employing a variety of biopolymers to coat different tablets, granules, beads, microspheres, etc.
(2) Biopolymeric matrix drug delivery systems composed of different non-swellable or swellable biopolymers.
(3) Slowly eroding drug delivery devices.
(4) Drug delivery devices controlled by the osmotic pressure.

Varieties of orally controlled drug-releasing systems have already been designed and developed for the releasing of numerous drug candidates over the prolonged period (Hasnain et al. 2016; Nayak et al. 2010c, d). In general, various shorter biological half-life drugs are required to be administered orally in multiple dosing to continue effectual plasma drug concentration level candidates over the prolonged period (Nayak 2011). So, there is a chance of occurrences of various systemic side effects. These negative aspects may be conquered by means of designing suitable controlled drug releasing dosage forms having the ability to transport drugs to the desired target sites in a controlled rate and also, for a longer time (Vyas and Khar 2006). Though, the challenge in the proper designing and optimal development of orally administrable drug delivery systems for controlled drug releasing is not just to produce sustained drug-releasing potential, but also to facilitate the prolong residence of the dosage in the GIT until the releasing of total drug(s) within the desired time span (Malakar et al. 2014a; Nayak et al. 2013a). Extended gastrointestinal residence augments the drug release rates as well as the drug-releasing duration. These lessen the drug wastages (Nayak et al. 2010c, d). In addition, these make possible the maintenance of a steady therapeutic level over a prolonged duration with the minimization of therapeutic level fluctuations (Malakar et al. 2014c). Orally controlled releasing drug delivery systems with prolonged gastrointestinal residence times are advantageous for drug candidates and these are (Patil et al. 2012):

(1) More effectual while the plasma drug levels are steady
(2) Drug absorption window in the gastrointestinal area
(3) Drug targeting facility at the sites in the upper gastrointestinal area
(4) Lower drug solubility at the higher pH.

1.3 Multiple Units as Drug-Releasing Carriers

Various orally administrable controlled drug-releasing systems can be classified as single-unit drug-releasing systems (Beg et al. 2012; Malakar et al. 2011; Nayak and Malakar 2010, 2011) and multiple-unit drug-releasing systems (Das et al. 2014; Hasnain and Nayak 2018b; Maji et al. 2012; Malakar and Nayak 2012a). During the past few decades, the designing of multiple-unit drug-releasing systems have been drawn the increasing considerations in the area of drug delivery research and technology (Bera et al. 2015a, b, c). Multiple-unit drug-releasing systems demonstrate several essential benefits over the single-unit drug releasing systems, because these systems have been shown the reduction of both inter-subject as well as intra-subject variabilities in the drug absorption process in addition to lowering the dose dumping possibilities (Elmowafy et al. 2009; Malakar and Nayak 2012a). At present, the designing and formulations of different multiple-unit drug-releasing systems such as nanoparticles, microparticles, spheroids, beads, particles, etc., is considered as one of the fashionable and thrust area of the drug delivery researches (Das et al. 2014; Guru et al. 2013; Jana et al. 2013b; Nayak and Pal. 2017a; Nayak et al. 2012, 2018c; Pal and Nayak 2011, 2017; Pawar et al. 2012). For the delivering of recommended total doses, multiple-units are sometimes capsulated in the capsules or filled into the sachets or compressed into the tablets. These multiple-units are capable of mixing with the gastrointestinal fluid to be distributed in wider region in the GIT that minimizes the impairment of drug-releasing performances owing to the malfunction of a small number of units (Elmowafy et al. 2009; Malakar and Nayak 2012a). These occurrences facilitate more predictable drug releasing facility. In addition, multiple-unit drug-releasing

Table 1.1 Rational for the use of multiple-unit orally administrable controlled drug-releasing systems over the single-unit systems

Characteristics	Multiple-unit systems	Single-unit systems
Gastric emptying	Uniform	Variable
Inter-subject and intra-subject variabilities in the drug absorption process	Low	High
Food effects (on the drug-releasing and pharmacokinetic actions)	Minimum	Significant influences on the integrity of dosage forms, mainly, for the hydrophilic matrices
Safety concerns because of dose dumping	Minimum	Significant (for drug candidates with narrow therapeutic index)
Drug loading	Often limited but enhanced with a technological application (e.g., extrusion)	High drug-loading possibility
Patient compliances	High (often beads in capsules) and less frequent intake	Could be an issue (if unit size is large)
Drug-releasing modulation	Customized profile possible	To some extent complicated

systems evade the vagaries of the gastric emptying event and various gastric transit rates; thus, the drugs are released more consistently with preventing the exposure to higher concentrations of drugs, when compared with the single-unit drug-releasing systems (Pal and Nayak 2011, 2017). Multiple-unit drug-releasing systems also lessen the possibilities of dose dumping occurrences and localized damaging of gastrointestinal mucosal surfaces (Elmowafy et al. 2009; Guru et al. 2013). The rationale for the use of multiple-unit orally administrable controlled drug-releasing systems over the single-unit systems is summarized in Table 1.1.

Recently, the exploitation of a variety of natural, semi-synthetic and synthetic biopolymers for the designing of different drug loaded multiple-unit oral administrable controlled drug-releasing systems such as nanoparticles, microparticles, spheroids, beads, particles, etc., have been the focal point of the current researches in the area of drug delivery (Malakar et al. 2013a; Nayak et al. 2011a, b; Pal and Nayak 2011). To design different kinds of multiple-unit orally administrable controlled drug releasing dosage forms, naturally derived biopolymers are currently being preferred over the synthetically derived biopolymers, as because these (i.e., naturally derived biopolymers) are freely extracted/obtained from the natural resources, less expensive, eco-friendly, nontoxic, and biodegradable (Nayak and Pal 2017; Nayak et al. 2018d). In the published literature, different kinds of multiple-unit orally administrable controlled drug-releasing dosage forms made of naturally derived biopolymers have been reported by various research groups (Nayak and Pal 2016, 2017a; Pal and Nayak 2017). All these reported multiple-unit drug-releasing systems for oral admin-

Table 1.2 Some examples of multiple-unit controlled drug-releasing dosage forms containing naturally derived biopolymers for oral administration

Multiple-unit controlled drug releasing dosage forms	Naturally derived biopolymers used	Drugs incorporated	References
Beads	Sodium alginate	Theophylline	Smrdel et al. (2008)
Beads	Sodium alginate	Diltiazem HCl	El-Kamel et al. (2003)
Beads	Sodium alginate	Metronidazole	Patel et al. (2006)
Beads	Sodium alginate	Gliclazide	Al-Kassas et al. (2007)
Beads	Sodium alginate	Sulindac	Yegin et al. (2007)
Beads	Sodium alginate	Ampicillin	Torre et al. (1998)
Beads	Sodium alginate	Diclofenac sodium	Morshad et al. (2012)
Microspheres	Sodium alginate	Furosemide	Das and Senapati (2008)
Beads	Pectin	Indomethacin	Chung and Zhang (2003)
Beads	Pectin	Rutin	Assifoui et al. (2011)
Beads	Pectin	Theophylline	Sriamornsak et al. (2010)
Beads	Pectin	Ketoprofen	El-Glibaly (2001)
Microbeads	Gellan gum	Amoxicillin	Babu et al. (2010)
Beads	Gellan gum	Amoxicillin trihydrate	Narkar et al. (2010)
Beads	Gellan gum	Cephalexin	Agnihotri et al. (2006)
Beads	Chitosan	Gliclazide	Varshosaz et al. (2009)
Beads	Chitosan	Ciprofloxacin	Srinatha et al. (2008)
Microspheres	Alginate–locust bean gum	Diclofenac sodium	Deshmukh et al. (2009b)
Microspheres	Alginate–pectinate	Aceclofenac	Chakraborty et al. (2010)
Beads	Alginate–gellan gum	Glipizide	Rao et al. (2007)
Microspheres	Alginate–gellan gum	Aceclofenac	Jana et al. (2013a)
Beads	Alginate–xanthan gum	Diclofenac sodium	Pongjayakul and Puttipipatkhachorn (2007)
Microspheres	Alginate–xanthan gum/guar gum/locust bean gum	Diclofenac sodium	Deshmukh et al. (2009a)

(continued)

Table 1.2 (continued)

Multiple-unit controlled drug releasing dosage forms	Naturally derived biopolymers used	Drugs incorporated	References
Beads	Alginate–gum Arabic	Glibenclamide	Nayak et al. (2012)
Beads	Sterculia gum–alginate	Pantoprazole	Singh et al. (2010)
Particles	Alginate–chitosan	Diclofenac sodium	González-Rodríguez et al. (2002)
Beads	Alginate–chitosan	Verapamil	Pasparakis and Bouropoulos (2006)
Beads	Alginate–chitosan	Carvedilol	Meng et al. (2011)
Beads	Alginate–chitosan	Ampicillin	Anal and Stevens (2005)
Microparticles	Alginate–chitosan	Prednisolone	Wittaya-areekul et al. (2006)
Beads	Alginate–chitosan	Nifedipine	Dai et al. (2008)
Particles	Pectinate–chitosan	Resveratrol	Das et al. (2011)
Beads	Pectinate–chitosan	Prednisone, Theophylline	Mennini et al. (2008)
Beads	Pectinate–high amylase starch	Diclofenac sodium	Desai (2005), Soares et al. (2013)
Beads	Gum Arabic–alginate	Glibenclamide	Nayak et al. (2012)
Microspheres	Tamarind gum–alginate	Gliclazide	Pal and Nayak (2012)
Beads	Tamarind gum–alginate	Metformin HCl	Nayak and Pal (2013c), Nayak et al. (2016)
Beads	Tamarind gum–pectinate	Metformin HCl	Nayak et al. (2014a)
Beads	Tamarind gum–alginate	Diclofenac sodium	Nayak and Pal (2011)
Beads	Tamarind gum–gellan gum	Metformin HCl	Nayak et al. (2014b)
Beads	Okra gum–alginate	Diclofenac sodium	Sinha et al. (2015)
Beads	Okra gum–alginate	Glibenclamide	Sinha et al. (2015)
Beads	Linseed polysaccharide–alginate	Diclofenac sodium	Hasnain et al. (2018b)
Beads	Fenugreek seed mucilage–alginate	Metformin HCl	Nayak et al. (2013)

(continued)

Table 1.2 (continued)

Multiple-unit controlled drug releasing dosage forms	Naturally derived biopolymers used	Drugs incorporated	References
Beads	Ispaghula husk mucilage–alginate	Glibenclamide	Nayak et al. (2013)
Beads	Ispaghula husk mucilage–alginate	Gliclazide	Nayak et al. (2010b)
Beads	Ispaghula husk mucilage–pectinate	Aceclofenac	Guru et al. (2018)
Beads	Ispaghula husk mucilage–pectinate	Metformin HCl	Nayak et al. (2014d)
Beads	Ispaghula husk mucilage–gellan gum	Metformin HCl	Nayak et al. (2014e)
Microparticles	Ispaghula husk mucilage–alginate	Isoniazid	Maurya et al. (2011)
Beads	Ispaghula husk mucilage–alginate	Metformin HCl	Sharma and Bhattacharya (2008)
Microbeads	Dellinia fruit gum–alginate	Timolol maleate	Sharma et al. (2009)
Microcapsules	Gum kondagogu–alginate	Glipizide	Krishna and Murthy (2010)
Beads	Fenugreek seed mucilage–alginate	Metformin HCl	Nayak et al. (2013)
Beads	Fenugreek seed mucilage–pectinate	Metformin HCl	Nayak et al. (2013d)
Beads	Fenugreek seed mucilage–gellan gum	Metformin HCl	Nayak and Pal (2014)
Microparticles	Chitosan–tamarind seed polysaccharide	Aceclofenac	Jana et al. (2013c)
Beads	Tapioca starch–alginate	Metoprolol tartrate	Biswas and Sahoo (2016)
Beads	Potato starch–alginate	Tolbutamide	Malakar et al. (2013b)
microbeads	Potato starch–alginate	Ibuprofen	Jha and Bhattacharya (2008)
Microbeads	Assam Bora rice starch–alginate	Metformin HCl	Sachan and Bhattyacharya (2006, 2009)
Beads	Jackfruit seed starch–alginate	Pioglitazone	Nayak et al. (2013e)

(continued)

Table 1.2 (continued)

Multiple-unit controlled drug releasing dosage forms	Naturally derived biopolymers used	Drugs incorporated	References
Beads	Jackfruit seed starch–pectinate	Metformin HCl	Nayak and Pal (2013a)
Beads	Jackfruit seed starch–alginate	Metformin HCl	Nayak and Pal (2013b)
Beads	Jackfruit seed starch–gellan gum	Metformin HCl	Nayak et al. (2014c)

istration were prepared by various techniques such as ionotropic gelation (Nayak and Pal 2013a; Nayak et al. 2011b, 2012, 2018e), covalent cross-linking gelation (Jana et al. 2013c), combined ionotropic/covalent cross-linking gelation (Bera et al. 2015b), ionotropic emulsification gelation (Guru et al. 2013; Nayak et al. 2013f), extrusion spheronization (Kulkarni et al. 2005), etc. Some examples of multiple-unit controlled drug-releasing dosage forms containing naturally derived biopolymers for oral administration are presented in Table 1.2.

References

S.A. Agnihotri, S.S. Jawalkar, T.M. Aminabhavi, Controlled release of cephalexin through gellan gum beads: Effect of formulation parameters on entrapment efficiency, size, and drug release. Eur. J. Pharm. Biopharm. **63**, 249–261 (2006)

R. Al-Kassas, O.M.N. Al-Gohary, M.M. Al-Fadhel, Controlling of systemic absorption of gliclazide through incorporation into alginate beads. Int. J. Pharm. **341**, 230–237 (2007)

A.K. Anal, W.F. Stevens, Chitosan-alginate multilayer beads for controlled release of ampicillin. Int. J. Pharm. **290**, 45–54 (2005)

T. Ansari, M.S. Hasnain, N. Hoda, A.K. Nayak, Microencapsulation of pharmaceuticals by solvent evaporation technique: a review. Elixir. Pharm. **47**, 8821–8827 (2012)

A. Assifoui, O. Chambin, P. Cayot, Drug release from calcium and zinc pectinate beads: Impact of dissolution medium composition. Carbohydr. Polym. **85**, 388–393 (2011)

R.J. Babu, S. Sathigrahi, M.T. Kumar, J.K. Pandit, Formulation of controlled release gellan gum macro beads of amoxicillin. Curr. Drug Deliv. **7**, 36–43 (2010)

S. Beg, A.K. Nayak, K. Kohli, S.K. Swain, M.S. Hasnain, Antibacterial activity assessment of a time-dependent release bilayer matrix tablet containing amoxicillin trihydrate. Brazilian J. Pharm. Sci. **48**, 265–272 (2012)

H. Bera, S. Boddupalli, S. Nandikonda, S. Kumar, A.K. Nayak, Alginate gel-coated oil-entrapped alginate–tamarind gum–magnesium stearate buoyant beads of risperidone. Int. J. Biol. Macromol. **78**, 102–111 (2015a)

H. Bera, S. Boddupalli, A.K. Nayak, Mucoadhesive-floating zinc-pectinate-sterculia gum interpenetrating polymer network beads encapsulating ziprasidone HCl. Carbohydr. Polym. **131**, 108–118 (2015b)

H. Bera, S.G. Kandukuri, A.K. Nayak, S. Boddupalli, Alginate-sterculia gum gel-coated oil-entrapped alginate beads for gastroretentive risperidone delivery. Carbohydr. Polym. **120**, 74–84 (2015c)

N. Biswas, R.K. Sahoo, Tapioca starch blended alginate mucoadhesive-floating beads for intragastric delivery of metoprolol tartrate. Int. J. Biol. Macromol. **83**, 61–70 (2016)

S. Chakraborty, M. Khandai, A. Sharma, N. Khanam, C.N. Patra, S.C. Dinda, K.K. Sen, Preparation, *in vitro* and *in vivo* evaluation of algino-pectinate bioadhesive microspheres: an investigation of the effects of polymers using multiple comparison analysis. Acta Pharm. **60**, 255–266 (2010)

Y.N. Chein, Controlled- and modulated-release drug delivery systems, in *Encyclopaedia of pharmaceutical technology*, ed. J. Swarbrick, J.C. Baylan, Marcel Dekker Inc.: New York, 1990), pp. 281–333

J.T. Chung, Z. Zhang, Mechanical characterization of calcium pectinate hydrogel for controlled drug delivery. Chem. Ind. **57**, 611–616 (2003)

Y.N. Dai, P. Lin, J.P. Zhang, A.Q. Wang, Q. Wei, Swelling characteristics and drug delivery properties of nifedipine-loaded pH-sensitive alginate-chitosan hydrogel beads. J. Biomed. Mater Res. Part B: Appl. Biomater **86B**, 493–500 (2008)

N.G. Das, S.K. Das, *Controlled-release of oral dosage forms*. Formul Fill & Finish 10–16 (2003)

M.K. Das, P.C. Senapati, Furosemide-loaded alginate microspheres prepared by ionic cross-linking technique: morphology and release characteristics. Indian J. Pharm. Sci. **70**, 77–84 (2008)

S. Das, A. Chaudhury, K.-Y. Ng, Preparation and evaluation of zinc-pectin-chitosan composite particles for drug delivery to the colon: Role of chitosan in modifying *in vitro* and *in vivo* drug release. Int. J. Pharm. **406**, 11–20 (2011)

B. Das, S. Dutta, A.K. Nayak, U. Nanda, Zinc alginate-carboxymethyl cashew gum microbeads for prolonged drug release: development and optimization. Int. J. Biol. Macromol. **70**, 505–515 (2014)

B. Das, S.O. Sen, R. Maji, A.K. Nayak, K.K. Sen, Transferosomal gel for transdermal delivery of risperidone. J. Drug. Deliv. Sci. Technol. **38**, 59–71 (2017)

V.N. Deshmukh, D.M. Sakarkar, R.D. Wakade, Formulation and evaluation of controlled release alginate microspheres using locust bean gum. J. Pharm. Res. **2**, 258–261 (2009a)

V.N. Deshmukh, J.K. Jadhav, V.J. Masirkar, D.M. Sakarkar, Formulation, optimization and evaluation of controlled release alginate microspheres using synergy gum blends. Res. J. Pharm. Tech. **2**, 324–327 (2009b)

K.G.H. Desai, Preparation and characteristics of high-amylose corn starch/pectin blend microparticles: a technical note. AAPS PharmSciTech 6, article 30 (2005)

I. El-Glibaly, Oral delayed-release system based on Zn-pectinate gel (ZPG) microparticles as an alternative carrier to calcium pectinate beads for colonic drug delivery. Int. J. Pharm. **232**, 199–221 (2001)

A.H. El-Kamel, O.M.N. Al-Gohary, E.A. Hosny, Alginate-diltiazem beads: optimization of formulation factors, *in vitro* and *in vivo* bioavailability. J. Microencapsul. **20**, 211–225 (2003)

E.M. Elmowafy, G.A.S. Awad, S. Mansour, A.E.-H.A. El-Shamy, Ionotropically emulsion gelled polysaccharide beads: Preparation, *in vitro* and *in vivo* evaluation. Carbohydr. Polym. **75**, 135–142 (2009)

D. Getsios, J.J. Caro, K.J. Ishak, W. El-Hadi, K. Payne, M. O'Connel, D. Albrecht, W. Feng, D. Dubois, Oxybutynin extended release and tolterodine immediate release: a health economic comparison. Clin. Drug. Invest. **24**, 81–88 (2004)

M.L. González-Rodríguez, M.A. Holgado, C. Sánchez-Lafuente, A.M. Rabasco, A. Fini, Alginate/chitosan particulate systems for sodium diclofenac release. Int. J. Pharm. **232**, 225–234 (2002)

P.R. Guru, H. Bera, M. Das, M.S. Hasnain, A.K. Nayak, Aceclofenac-loaded *Plantago ovata* F. husk mucilage-Zn^{+2}-pectinate controlled-release matrices. Starch-Stärke **70**, 1700136 (2018)

P.R. Guru, A.K. Nayak, R.K. Sahu, Oil-entrapped sterculia gum-alginate buoyant systems of aceclofenac: development and *in vitro* evaluation. Colloids Surf. B: Biointerf. **104**, 268–275 (2013)

M.S. Hasnain, A.K. Nayak, Alginate-inorganic composite particles as sustained drug delivery matrices, in *Applications of Nanocomposite materials in drug delivery*, ed. A.A.M. Inamuddin, A. Mohammad (A volume in Woodhead Publishing Series in Biomaterials, Elsevier Inc., 2018a), pp. 39–74

M.S. Hasnain, A.K. Nayak, Chitosan as responsive polymer for drug delivery applications, in *Stimuli responsive polymeric Nanocarriers for drug delivery applications*, ed. A.S.H. Makhlouf, Abu-Thabit N.Y. (Volume 1, Types and Triggers, Woodhead Publishing Series in Biomaterials, Elsevier Ltd., 2018b), pp. 581–605

M.S. Hasnain, A.K. Nayak, Nanocomposites for improved orthopedic and bone tissue engineering applications, in *Applications of Nanocomposite materials in orthopedics*, ed. A.A.M. Inamuddin, A. Mohammad (A volume in Woodhead Publishing Series in Biomaterials, Elsevier Inc., 2019), pp. 145–177

M.S. Hasnain, A.K. Nayak, M. Singh, M. Tabish, M.T. Ansari, T.J. Ara, Alginate-based bipolymeric-nanobioceramic composite matrices for sustained drug release. Int. J. Biol. Macromol. **83**, 71–77 (2016)

M.S. Hasnain, P. Rishishwar, S. Rishishwar, S. Ali, A.K. Nayak, Isolation and characterization of *Linum usitatisimum* polysaccharide to prepare mucoadhesive beads of diclofenac sodium. Int. J. Biol. Macromol. **116**, 162–172 (2018)

J.J. Hirani, D.A. Rathod, K.R. Vadalia, Orally disintegrating tablets: a review. Trop. J. Pharm. Res. **8**, 161–172 (2009)

S. Jana, N. Maji, A.K. Nayak, K.K. Sen, S.K. Basu, Development of chitosan-based nanoparticles through inter-polymeric complexation for oral drug delivery. Carbohydr. Polym. **98**, 870–876 (2013a)

S. Jana, A. Saha, A.K. Nayak, K.K. Sen, S.K. Basu, Aceclofenac-loaded chitosan-tamarind seed polysaccharide interpenetrating polymeric network microparticles. Colloids Surf. B: Biointerf. **105**, 303–309 (2013b)

S. Jana, S.A. Ali, A.K. Nayak, K.K. Sen, S.K. Basu, Development and optimization of topical gel containing aceclofenac-crospovidone solid dispersion by "Quality by Design" approach. Chem. Eng. Res. Des. **92**, 2095–2105 (2014)

S. Jana, A. Gangopadhaya, B.B. Bhowmik, A.K. Nayak, A. Mukhrjee, Pharmacokinetic evaluation of testosterone-loaded nanocapsules in rats. Int. J. Biol. Macromol. **72**, 28–30 (2015a)

S. Jana, A. Samanta, A.K. Nayak, K.K. Sen, S. Jana, Novel alginate hydrogel core–shell systems for combination delivery of ranitidine HCl and aceclofenac. Int. J. Biol. Macromol. **74**, 85–92 (2015b)

A.K. Jha, A. Bhattacharya, Preparation and *in vitro* evaluation of sweet potato starch blended sodium alginate microbeads. Adv. Nat. Appl. Sci. **2**, 122–128 (2008)

R.R. Krishna, T.E.G.K. Murthy, Preparation and evaluation of mucoadhesive microcapsules of glipizide formulated with gum: *In vitro* and *in vivo*. Acta. Pharm. Sci. **52**, 335–344 (2010)

G.T. Kulkarni, K. Gowthamarajan, R.R. Dhobe, F. Yohanan, B. Suresh, Development of controlled release spheroids using natural polysaccharide as release modifier. Drug. Deliv. **12**, 201–206 (2005)

M.A. Longer, J.R. Robinson, Sustained-release drug delivery systems, in *Remington's pharmaceutical sciences*, 18th edn., ed. by A.R. Gennaro (Mark Easton Publishing Company, New York, 1990), p. 1676

J. Malakar, A.K. Nayak, Theophylline release behavior for hard gelatin capsules containing various hydrophilic polymers. J. Pharm. Educ. Res. **3**, 10–16 (2012a)

J. Malakar, A.K. Nayak, Formulation and statistical optimization of multiple-unit ibuprofen-loaded buoyant system using 2^3-factorial design. Chem. Eng. Res. Des. **9**, 1834–1846 (2012b)

J. Malakar, A.K. Nayak, Floating bioadhesive matrix tablets of ondansetron HCl: optimization of hydrophilic polymer-blends. Asian J. Pharm. **7**, 174–183 (2013)

J. Malakar, P.K. Datta, S. Dey, A. Gangopadhyay, A.K. Nayak, Proniosomes: a preferable carrier for drug delivery system. Elixir. Pharm. **40**, 5120–5124 (2011)

J. Malakar, A. Gangopadhyay, A.K. Nayak, Transferosome: an opportunistic carrier for transdermal drug delivery system. Int. Res. J. Pharm. **3**, 35–38 (2012a)

J. Malakar, A. Ghosh, A. Basu, A.K. Nayak, Nanotechnology: a promising carrier for intracellular drug delivery system. Int. Res. J. Pharm. **3**, 36–40 (2012b)

J. Malakar, S.O. Sen, A.K. Nayak, K.K. Sen, Formulation, optimization and evaluation of transferosomal gel for transdermal insulin delivery. Saudi. Pharm. J. **20**, 355–363 (2012c)

J. Malakar, A.K. Nayak, D. Pal, Development of cloxacillin loaded multiple-unit alginate-based floating system by emulsion–gelation method. Int. J. Biol. Macromol. **50**, 138–147 (2012d)

J. Malakar, A.K. Nayak, A. Das, Modified starch (cationized)-alginate beads containing aceclofenac: formulation optimization using central composite design. Starch-Stärke. **65**, 603–612 (2013a)

J. Malakar, A.K. Nayak, P. Jana, D. Pal, Potato starch-blended alginate beads for prolonged release of tolbutamide: development by statistical optimization and *in vitro* characterization. Asian J. Pharm. **7**, 43–51 (2013b)

J. Malakar, A. Basu, A.K. Nayak, Candesartan cilexetil microemulsions for transdermal delivery: Formulation, *in-vitro* skin permeation and stability assessment. Curr. Drug Deliv. **11**, 313–321 (2014a)

J. Malakar, P. Dutta, S.D. Purokayastha, S. Dey, A.K. Nayak, Floating capsules containing alginate-based beads of salbutamol sulfate: *in vitro-in vivo* evaluations. Int. J. Biol. Macromol. **64**, 181–189 (2014b)

J. Malakar, K. Das, A.K. Nayak, *In situ* cross-linked matrix tablets for sustained salbutamol sulfate release - formulation development by statistical optimization. Polym. Med. **44**, 221–230 (2014c)

D.P. Maurya, Y. Sultana, M. Aquil, D. Kumar, K. Chuttani, A. Ali, A.K. Mishra, Formulation and optimization of alkaline extracted ispaghula husk microparticles of isoniazid – *in vitro* and *in vivo* assessment. J. Microencapsul. **28**, 472–482 (2011)

X. Meng, P. Li, Q. Wei, H.-X. Zhang, pH sensitive alginate-chitosan hydrogel beads for carvedilol delivery. Pharm. Dev. Technol. **16**, 22–28 (2011)

N. Mennini, S. Furlanetto, F. Maestrelli, S. Pinzauti, P. Mura, Response surface methodology in the optimization of chitosan-Ca pectinate bead formulations. Eur. J. Pharm. Sci. **35**, 318–325 (2008)

M.M. Morshad, J. Mallick, A.K. Nath, M.Z. Uddin, M. Dutta, M.A. Hossain, M.H. Kawsar, Effect of barium chloride as a cross-linking agent on the sodium alginate based diclofenac sodium beads. Bangladesh. Pharm. J. **15**, 53–57 (2012)

J. Malakar, S.O. Sen, A.K. Nayak, K.K. Sen, Development and evaluation of microemulsion for transdermal delivery of insulin. ISRN Pharm 2011, Article ID 780150 (2011)

R.H. Muller, C. Jacobs, O. Kayser, Nanosuspensions as particulate drug formulation therapy: rational for development and what we can expect for the future. Adv. Drug Deliv. Rev. **47**, 3–19 (2001)

R. Maji, B. Das, A.K. Nayak, S. Ray, Ethyl cellulose microparticles containing metformin HCl by emulsification-solvent evaporation technique: effect of formulation variables. ISRN Polym Sci 2012, Article ID 801827 (2012)

M. Narkar, P. Sher, A. Pawar, Stomach-specific controlled release gellan beads of acid-soluble drug prepared by ionotropic gelation method. AAPS PharmSciTech. **11**, 267–277 (2010)

A.K. Nayak, Advances in therapeutic protein production and delivery. Int. J. Pharm. Pharmaceut. Sci. **2**, 1–5 (2010)

A.K. Nayak, Controlled release drug delivery systems. Sci. J. UBU **2**, 1–8 (2011)

A.K. Nayak, S. Jana, A solid self-emulsifying system for dissolution enhancement of etoricoxib. J. PharmSciTech. **2**, 87–90 (2013)

A.K. Nayak, J. Malakar, Formulation and *in vitro* evaluation of gastroretentive hydrodynamically balanced system for ciprofloxacin HCl. J. Pharm. Educ. Res. **1**, 65–68 (2010)

A.K. Nayak, J. Malakar, Formulation and *in vitro* evaluation of hydrodynamically balanced system for theophylline delivery. J. Basic Clin. Pharm. **2**, 133–137 (2011)

A.K. Nayak, K. Manna, Current developments in orally disintegrating tablet technology. J. Pharm. Educ. Res. **2**, 24–38 (2011)

A.K. Nayak, D. Pal, Development of pH-sensitive tamarind seed polysaccharide-alginate composite beads for controlled diclofenac sodium delivery using response surface methodology. Int. J. Biol. Macromol. **49**, 784–793 (2011)

A.K. Nayak, D. Pal, Blends of jackfruit seed starch-pectin in the development of mucoadhesive beads containing metformin HCl. Int. J. Biol. Macromol. **62**, 137–145 (2013a)

A.K. Nayak, D. Pal, Ionotropically-gelled mucoadhesive beads for oral metformin HCl delivery: formulation, optimization and antidiabetic evaluation. J. Sci. Ind. Res. **72**, 15–22 (2013b)

A.K. Nayak, D. Pal, Formulation optimization of jackfruit seed starch-alginate mucoadhesive beads of metformin HCl. Int. J. Biol. Macromol. **59**, 264–272 (2013c)

A.K. Nayak, D. Pal, Chitosan-based interpenetrating polymeric network systems for sustained drug release, in *Advanced theranostics materials*, ed. by A. Tiwari, H.K. Patra, J.-W. Choi (WILEY-Scrivener, USA, 2015), pp. 183–208

A.K. Nayak, D. Pal, Sterculia gum-based hydrogels for drug delivery applications, in *Polymeric Hydrogels as smart Biomaterials*, ed. by S. Kalia (Springer Series on Polymer and Composite Materials, Springer International Publishing, Switzerland, 2016), pp. 105–151

A.K. Nayak, K.K. Sen, Hydroxyapatite-ciprofloxacin implantable minipellets for bone delivery: Preparation, characterization, *in vitro* drug adsorption and dissolution studies. Int. J. Drug. Dev. Res. **1**, 47–59 (2009)

A.K. Nayak, A. Bhattacharya, K.K. Sen, Hydroxyapatite-antibiotic implantable minipellets for bacterial bone infections using precipitation technique: Preparation, characterization and *in-vitro* antibiotic release studies. J. Pharm. Res. **3**, 53–59 (2010a)

A.K. Nayak, B. Mohanty, K.K. Sen, Comparative evaluation of *in vitro* diclofenac sodium permeability across excised mouse skin from different common pharmaceutical vehicles. Int. J. PharmTech Res. **2**, 920–930 (2010b)

A.K. Nayak, R. Maji, B. Das, Gastroretentive drug delivery systems: a review. Asian J. Pharm. Clin. Res. **3**, 2–10 (2010c)

A.K. Nayak, J. Malakar, K.K. Sen, Gastroretentive drug delivery technologies: current approaches and future potential. J. Pharm. Educ. Res. **1**, 1–12 (2010d)

A.K. Nayak, A. Bhattacharyya, K.K. Sen, *In vivo* ciprofloxacin release from hydroxyapatite-ciprofloxacin bone-implants in rabbit tibia. ISRN Orthop 2011, Article ID 420549 (2011a)

A.K. Nayak, B. Laha, K.K. Sen, Development of hydroxyapatite-ciprofloxacin bone-implants using "Quality by Design". Acta Pharm. **61**, 25–36 (2011b)

A.K. Nayak, S. Khatua, M.S. Hasnain, K.K. Sen, Development of alginate-PVP K 30 microbeads for controlled diclofenac sodium delivery using central composite design. DARU J. Pharm. Sci. **19**, 356–366 (2011c)

A.K. Nayak, B. Das, R. Maji, Calcium alginate/gum Arabic beads containing glibenclamide: Development and *in vitro* characterization. Int. J. Biol. Macromol. **51**, 1070–1078 (2012)

A.K. Nayak, M.S. Hasnain, J. Malakar, Development and optimization of hydroxyapatite-ofloxacin implants for possible bone-implantable delivery in osteomyelitis treatment. Curr. Drug Deliv. **10**, 241–250 (2013a)

A.K. Nayak, B. Das, R. Maji, Gastroretentive hydrodynamically balanced system of ofloxacin: formulation and *in vitro* evaluation. Saudi. Pharm. J. **21**, 113–117 (2013b)

A.K. Nayak, S. Kalia, M.S. Hasnain, Optimization of aceclofenac-loaded pectinate-poly (vinyl pyrrolidone) beads by response surface methodology. Int. J. Biol. Macromol. **62**, 194–202 (2013c)

A.K. Nayak, D. Pal, J. Malakar, Development, optimization and evaluation of emulsion-gelled floating beads using natural polysaccharide-blend for controlled drug release. Polym. Eng. Sci. **53**, 338–350 (2013e)

A.K. Nayak, D. Pal, J. Pradhan, M.S. Hasnain, Fenugreek seed mucilage-alginate mucoadhesive beads of metformin HCl: design, optimization and evaluation. Int. J. Biol. Macromol. **54**, 144–154 (2013f)

A.K. Nayak, D. Pal, K. Santra, Tamarind seed polysaccharide-gellan mucoadhesive beads for controlled release of metformin HCl. Carbohydr. Polym. **103**, 154–163 (2014a)

A.K. Nayak, D. Pal, K. Santra, Development of pectinate-ispagula mucilage mucoadhesive beads of metformin HCl by central composite design. Int. J. Biol. Macromol. **66**, 203–221 (2014b)

A.K. Nayak, D. Pal, K. Santra, Ispaghula mucilage-gellan mucoadhesive beads of metformin HCl: development by response surface methodology. Carbohydr. Polym. **107**, 41–50 (2014c)

A.K. Nayak, D. Pal, K. Santra, *Artocarpus heterophyllus* L. seed starch-blended gellan gum mucoadhesive beads of metformin HCl. Int. J. Biol. Macromol. **65**, 329–339 (2014d)

A.K. Nayak, D. Pal, K. Santra, Swelling and drug release behavior of metformin HCl-loaded tamarind seed polysaccharide-alginate beads. Int. J. Biol. Macromol. **82**, 1023–1027 (2016)

A.K. Nayak, D. Pal, K. Santra, Development of calcium pectinate-tamarind seed polysaccharide mucoadhesive beads containing metformin HCl. Carbohydr. Polym. **101**, 220–230 (2014a)

A.K. Nayak, D. Pal, K. Santra, *Plantago ovata* F. Mucilage-alginate mucoadhesive beads for controlled release of glibenclamide: development, optimization, and *in vitro-in vivo* evaluation. J. Pharm. 2013, Article ID 151035 (2013g)

A.K. Nayak, D. Pal, Natural starches-blended ionotropically-gelled micrparticles/beads for sustained drug release, in *Handbook of composites from renewable materials*, ed. V.K. Thakur, M.K. Thakur, M.R. Kessler (Volume 8, Nanocomposites: Advanced Applications, Wiley-Scrivener, USA 2017), pp. 527–560

A.K. Nayak, K.K. Sen, Bone-targeted drug delivery systems, in *Bio-Targets & Drug Delivery Approaches*, ed. S. Maiti, K.K. Sen (CRC Press, 2016), pp. 207–231

A.K. Nayak, A.A. Syed, S. Beg, T.J. Ara, M.S. Hasnain, Drug delivery: present, past and future of medicine, in *Applications of Nanocomposite materials in drug delivery*, ed. A.A.M. Inamuddin, A. Mohammad (A volume in Woodhead Publishing Series in Biomaterials, Elsevier Inc., 2018a) 255–282

A.K. Nayak, S. Beg, M.S. Hasnain, J. Malakar, D. Pal, Soluble starch-blended Ca^{2+}-Zn^{2+}-alginate composites-based microparticles of aceclofenac: formulation development and *in vitro* characterization. Future J. Pharm. Sci. **4**, 63–70 (2018b)

A.K. Nayak, B. Das, Introduction to polymeric gels, in *Polymeric gels characterization, properties and biomedical applications*, ed. K. Pal, I. Bannerjee (A volume in Woodhead Publishing Series in Biomaterials, Elsevier Ltd., 2018c), pp. 3–27

A.K. Nayak, T.J. Ara, M.S. Hasnain, N. Hoda, Okra gum-alginate composites for controlled releasing drug delivery, in *Applications of Nanocomposite materials in drug delivery*, ed. A.A.M. Inamuddin, A. Mohammad (A volume in Woodhead Publishing Series in Biomaterials, Elsevier Inc., 2018d), pp. 761–785

A.K. Nayak, M.S. Hasnain, D. Pal, Gelled microparticles/beads of sterculia gum and tamarind gum for sustained drug release, in *Handbook of springer on polymeric gel*, ed. V.K. Thakur, M.K. Thakur (Springer International Publishing, Switzerland, 2018e), pp. 361–414

D. Pal, A.K. Nayak, Alginates, blends and microspheres: controlled drug delivery, in *Encyclopedia of biomedical polymers and polymeric biomaterials,* ed. M. Mishra (Taylor & Francis Group, New York, NY 10017, U.S.A., Vol. I, 2015a), pp. 89–98

D. Pal, A.K. Nayak, Interpenetrating polymer networks (IPNs): Natural polymeric blends for drug delivery, in *Encyclopedia of biomedical polymers and polymeric biomaterials*, ed. M. Mishra (Taylor & Francis Group, New York, NY 10017, U.S.A., Vol. VI, 2015b), pp. 4120–4130

Y.L. Patel, P. Sher, A.P. Pawar, The effect of drug concentration and curing time on processing and properties of calcium alginate beads containing metronidazole by response surface methodology. AAPS PharmSciTech 7, Article 86 (2006)

D. Pal, A.K. Nayak, Plant polysaccharides-blended ionotropically-gelled alginate multiple-unit systems for sustained drug release, in *Handbook of composites from renewable materials*, ed. V.K. Thakur, M.K. Thakur, M.R. Kessler (Volume 6, Polymeric Composites, WILEY-Scrivener, USA, 2017), pp. 399–400

D. Pal, A.K. Nayak, Development, optimization and anti-diabetic activity of gliclazide-loaded alginate-methyl cellulose mucoadhesive microcapsules. AAPS PharmSciTech. **12**, 1431–1441 (2011)

D. Pal, A.K. Nayak, Novel tamarind seed polysaccharide-alginate mucoadhesive microspheres for oral gliclazide delivery. Drug. Deliv. **19**, 123–131 (2012)

D. Pal, A.K. Nayak, S. Saha, Interpenetrating polymer network hydrogels of chitosan: applications in controlling drug release, in *Cellulose-based superabsorbent hydrogels, polymers and polymeric composites: a reference series*, ed. by I.H. Mondal (Springer, Cham, 2018), pp. 1–41

G. Pasparakis, N. Bouropoulos, Swelling studies and *in vitro* release of verapamil from calcium alginate-chitosan beads. Int. J. Pharm. **323**, 34–42 (2006)

P. Patil, D. Chavanke, M. Wagh, A review on ionotropic gelation method: Novel approach for controlled gastroretentive gelispheres. Int. J. Pharm. Pharmaceut. Sci. **4**, 27–32 (2012)

H. Pawar, D. Douroumis, J.S. Boateng, Preparation and optimization of PMMA-chitosan-PEG nanoparticles for oral drug delivery. Colloids Surf. B: Biointerf. **90**, 102–108 (2012)

T. Pongjayakul, S. Puttipipatkhachorn, Xanthan-alginate composite gel beads: molecular interaction and *in vivo* characterization. Int. J. Pharm. **331**, 61–71 (2007)

V.U. Rao, M. Vasudha, K. Bindu, S. Samanta, P.S. Rajinikanth, B. Mishra, J. Balasubhramaniam, Formulation and *in vitro* characterization of sodium alginate-gellan beads of glipizide. Acta Pharm. Sin. **49**, 13–28 (2007)

S.N. Rath Adhikari, B.S. Nayak, A.K. Nayak, B. Mohanty, Formulation and evaluation of buccal patches for delivery of atenolol. AAPS PharmSciTech **11**, 1034–1044 (2010)

S. Ray, P. Sinha, B. Laha, S. Maiti, U.K. Bhattacharyya, A.K. Nayak, Polysorbate 80 coated crosslinked chitosan nanoparticles of ropinirole hydrochloride for brain targeting. J. Drug. Deliv. Sci. Technol. **48**, 21–29 (2018)

N.K. Sachan, A. Bhattyacharya, Evaluation of Assam Bora rice starch as a possible natural mucoadhesive polymer in the formulation of microparticulate controlled drug delivery systems. J. Assam. Sci. Soc. **47**, 34–41 (2006)

N.K. Sachan, A. Bhattyacharya, Feasibility of Assam bora rice based matrix microdevices for controlled release of water insoluble drug. Int. J. Pharm. Pharmaceut. Sci. **1**, 96–102 (2009)

V.K. Sharma, A. Bhattacharya, Release of metformin hydrochloride from ispaghula-sodium alginate beads adhered on cock intestinal mucosa. Indian J. Pharm. Educ. Res. **42**, 365–372 (2008)

H.K. Sharma, B. Sarangi, S.P. Pradhan, Preparation and *in vitro* evaluation of mucoadhesive microbeads containing timolol maleate using mucoadhesive substances of *Dellinia india* L. Arch. Pharm. Res. **1**, 181–18 (2009)

J. Siepmann, F. Siepmann, The modified-release drug delivery landscape: academic viewpoint, in *Modified release drug delivery technology*, 2nd edn., ed. by M.J. Rathbone, J. Hadgraft, M.S. Roberts, M.E. Lane (Informa Healthcare Inc., USA, New York, 2008), pp. 17–34

B. Singh, V. Sharma, D. Chauhan, Gastroretentive floating sterculia-alginate beads for use in antiulcer drug delivery. Chem. Eng. Res. Des. **88**, 997–1012 (2010)

P. Sinha, U. Ubaidulla, M.S. Hasnain, A.K. Nayak, B. Rama, Alginate-okra gum blend beads of diclofenac sodium from aqueous template using $ZnSO_4$ as a cross-linker. Int. J. Biol. Macromol. **79**, 555–563 (2015)

P. Smrdel, M. Bogataj, A. Mrhar, The influence of selected parameters on the size and shape of alginate beads prepared by ionotropic gelation. Sci. Pharm. **76**, 77–89 (2008)

G.A. Soares, A.D. de Castro, B.S.F. Cury, R.C. Evangelista, Blends of cross-linked high amylase starch/pectin loaded with diclofenac. Carbohydr. Polym. **91**, 135–142 (2013)

P. Sriamornsak, J. Nunthanid, K. Cheewatanakornkool, S. Manchun, Effect of drug loading method on drug content and drug release from calcium pectinate gel beads. AAPS PharmSciTech. **11**, 1315–1319 (2010)

A. Srinatha, J.K. Pandit, S. Singh, Ionic cross-linked chitosan beads for extended release of ciprofloxacin: *in vitro* characterization. Indian J. Pharm. Sci. **70**, 16–21 (2008)

R. Tiwari, G. Tiwari, A.K. Rai, Controlled drug delivery. Indian Pharmacist **8**, 29–34 (2009)

M.L. Torre, P. Giunchedi, L. Maggi, R. Stefli, E.O. Mechiste, U. Conte, Formulation and characterization of calcium alginate beads containing ampicillin. Pharm. Dev. Technol. **3**, 193–198 (1998)

J. Varshosaz, N. Tavakoli, M. Minayian, N. Rahdri, Applying the Taguchi design for optimized formulation of sustained release gliclazide chitosan beads. An *in vitro/in vivo* study. AAPS PharmSciTech. **10**, 158–165 (2009)

M.V.S. Verma, A.M. Kaushal, A. Garg, S. Garg, Factors affecting mechanism and kinetics of drug release from matrix-based oral controlled drug delivery systems. Am. J. Drug. Deliv. **2**, 43–57 (2004)

A. Verma, J. Dubey, N. Verma, A.K. Nayak, Chitosan-hydroxypropyl methylcellulose matrices as carriers for hydrodynamically balanced capsules of moxifloxacin HCl. Curr. Drug Deliv. **14**, 83–90 (2017)

S.P. Vyas, R.K. Khar, Essentials of controlled drug delivery, in *Controlled drug delivery*. Concepts and advances (1st edn, Delhi, India, Vallabh Prakashan, 2006), pp. 1–53

S. Wittaya-areekul, J. Kruenate, C. Prahsarn, Preparation and *in vitro* evaluation of mucoadhesive properties of alginate/chitosan microparticles containing prednisolone. Int. J. Pharm. **312**, 113–118 (2006)

B.A. Yegin, B. Moulari, N.T. Durlu-Kandilci, P. Korkusuz, Y. Pellequer, A. Lamprecht, Sulindac loaded alginate beads for a mucoprotective and controlled drug release. J. Microencapsul. **24**, 371–382 (2007)

Chapter 2
Plant Polysaccharides in Drug Delivery Applications

2.1 Natural Polysaccharides

Nowadays, the entire world is moving en route for the utilization of natural excipients in diverse biomedical applications (Hasnain et al. 2010; Mano et al. 2007; Nayak 2010; Nayak and Pal 2012, 2017a). Excipients of natural origin do hold preferences over the excipients synthesized by chemical routes (i.e. synthetic excipients), as these are nontoxic, biologically degradable and uninhibitedly accessible from the naturally occurring origins (Pal et al. 2010; Nayak and Pal 2012; Nayak et al. 2010). With the enhanced requirement of these naturally occurring substances, it has been important for investigating more new resources to meet up considerable technological and industrial requirements. At the present time, the socioeconomic circumstances of the modern world have raised the interest of naturally derived biomaterials particularly plant-derived materials in biomedical applications. The naturally derived polymers are biocompatible, biodegradable and economic (Pal and Nayak 2015a, b; Nayak and Pal 2018). Usually, lower extraction expenditure of natural polymers interrelated to the huge accessibility from the natural resources is one of the important additional advantages (Malafaya et al. 2007; Nayak et al. 2018a, b). Actually, natural polymers compose structurally different categories of biological macromolecules with an extensive range of physicochemical characteristics (Hasnain et al. 2010; Malafaya et al. 2007; Mano et al. 2007). In the biomedical field, the natural polymers are degraded into various physiological metabolites, which make them as excellent excipient biopolymeric candidates for different biomedical uses including drug delivery (Hasnain and Nayak 2018b; Nayak and Pal 2017a). Natural polymers may also include extracellular materials (called as ligands), which are essential for binding with the cell receptors and employed in the targeted delivery of drugs and biologicals (Pal and Nayak 2017). Various natural polymers can be classified on the basis of their constituents (Hasnain et al. 2010):

© The Author(s), under exclusive license to Springer Nature Singapore Pte Ltd. 2019
A. K. Nayak and M. S. Hasnain, *Plant Polysaccharides-Based Multiple-Unit Systems for Oral Drug Delivery*, SpringerBriefs in Applied Sciences and Technology, https://doi.org/10.1007/978-981-10-6784-6_2

(i)   Protein-based natural polymers:

    (1)   Gelatin,
    (2)   Collagen,
    (3)   Albumin,
    (4)   Fibrin,
    (5)   Silk fibroin, etc.

(ii) Natural polysaccharides (carbohydrates):

    (1)   Guar gum,
    (2)   Gum tragacanth,
    (3)   Gum acacia,
    (4)   Starches,
    (5)   Pectin,
    (6)   Okra gum,
    (7)   Tamarind gum,
    (8)   Alginates,
    (9)   Gellan gum,
    (10) Chitin and chitosan,
    (11) Xanthan gum,
    (12) Sterculia gum,
    (13) Locust bean gum, etc.

Based on the origin (source), natural polymers are categorized as follows (Malafaya et al. 2007; Mano et al. 2007):

(1)   Animal origin: e.g., chitin, chitosan, chondroitin, hyaluronic acid, etc.
(2)   Plant origin gum: e.g., gum Arabica, pectin, cashew gum, guar gum, tamarind gum, okra gum, starches, etc.
(3)   Algal origin gum: e.g., alginates, agar-agar, etc.
(4)   Microbial origin gum: e.g., dextran, gellan gum, xanthan gum, etc.

2.2   Plant Polysaccharides as Pharmaceutical Excipients

Amongst numerous sources of naturally occurring substances, the plant resources are considered as a potentially renewable resource group (Hati et al. 2014; Jena et al. 2012a, b; Nayak and Pal 2012, 2017a; Sinha Mahapatra et al. 2011). If these plant resources are being cultivated and/or harvested in an effective sustainable mode, these can offer a constant volume of supplying a huge volume of plant-derived natural polymers (Pal et al. 2010; Nayak and Pal 2017a, b). Almost all the plant polysaccharides occur in the forms of gums, mucilages, and starches within various plant parts like leaves, stems, roots, rhizomes, corms, cereals, barks, exudates, etc. (Nayak and Pal 2017a, b; Nayak et al. 2018a). The molecular structures of these plant polysaccharides possess complex and branched polysaccharide structures, which

Table 2.1 Sources of some important plant polysaccharides for the use as drug delivery excipients

Plant polysaccharides	Sources		Plant polysaccharides
	Botanical names	Family	
Gum Arabica	*Acacia arabica*	Leguminoseae	Nayak et al. (2012a)
Guar gum	*Cyamompsis tetraganolobus*	Leguminoseae	Krishnaiah et al. (2001)
Sterculia (karaya) gum	*Sterculia urens*	Sterculiaceae	Guru et al. (2012)
Locust bean gum	*Ceratonia siliqua*	Fabaceae	Deshmukh et al. (2009b)
Tamarind gum	*Tamarindus indica*	Leguminoseae	Nayak et al. (2014a, b)
Konjac glucomannan	*Amorphophallus konjac*	Araceae	Fan et al. (2008)
Okra gum	*Hibiscus esculantus*	Malvaceae	Sinha et al. (2015a, b)
Khaya gum	*Khaya grandifoliola*	Meliaceae	Odeku and Fell (2004)
Gum kondagogu	*Cochlospermum gossypium*	Colchospermaceae	Malik et al. (2012)
Cashew tree gum	*Anacardium occidentale*	Anacardiaceae	Hasnain et al. (2017a, b, 2018a)
Moringa gum	*Moringa oleifera*	Moringaceae	Panda et al. (2008)
Abelmoschus gum	*Abelmoschus esculantus*	Malvaceae	Ofoefule and Chukwu (2001)
Albizia gum	*Albizia procera*	Leguminoseae	Pachuau and Mazumdar (2012)
Gum odina	*Lannea woodier*	Anacardiaceae	Sinha et al. (2011)
Terminalia gum	*Terminalia randii*	Combretaceae	Bamiro et al. (2012)
Gum cordia	*Cordia obliqua*	Boraginaceae	Ahuja et al. (2010)
Dillenia fruit gum	*Dillenia indica* L.	Dilleniaceae	Ketousetuo and Bandyopadhyay (2007)
Linseed polysaccharide	*Linum usitatissimum* L.,	Linaceae	Hasnain et al. (2018b)
Aloe mucilage	*Aloe barbadensis*	Liliaceae	Jani et al. (2007)
Ispaghula mucilage	*Plantago ovata*	Plantaginaceae	Nayak et al. (2013h)
Mimosa pudica seed mucilage	*Mimosa pudica*	Mimosaceae	Singh et al. (2001)
Fenugreek seed mucilage	*Trigonella foenum-graecum* L.	Fabaceae	Nayak et al. (2012b)

(continued)

Table 2.1 (continued)

Plant polysaccharides	Sources		Plant polysaccharides
	Botanical names	Family	
Spinacia oleraceae L. leaves mucilage	*Spinacia oleraceae* L.	Amaranthaceae	Nayak et al. (2010f)
Basella alba L. leaves mucilage	*Basella alba* L.	Basellaceae	Pal et al. (2010)
Jackfruit seed starch	*Artocarpus heterophyllus*	Moraceae	Nayak and Pal (2013a, b, 2014c)
Potato starch	*Solanum tuberosum* L.	Solanaceae	Malakar et al. (2013b)

consist of multiple sugar units linked together (Nayak et al. 2015; Pal and Nayak 2017). Most of the plant polysaccharides are used in the food industry and also regarded as safe for human consumption (Beneke et al. 2009; Avachat et al. 2011).

During the past few years, numerous plant polysaccharides are being explored exploited as potential drug delivery excipients in the formulations of various drug delivery systems (Beneke et al. 2009; Pal and Nayak 2017). These recently explored plant polysaccharides have been studied as drug delivery excipients in the formulation of different pharmaceutical drug delivery dosage forms such as tablets, capsules, nanoparticles, microparticles, beads, spheroids, pellets, creams, gels, emulsions, suspensions, pastes, transdermal patches, buccal patches, etc. (Avachat et al. 2011). These plant polysaccharide drug delivery excipients have also been utilized as tablet binders, tablet disintegrants, emulsifiers, suspending agents, gelling agents, film formers, bio- or mucoadhesive agents, enteric resistants, matrix formers, release retardants, etc., in different pharmaceutical drug delivery dosage forms (Jani et al. 2007; Beneke et al. 2009; Avachat et al. 2011). The sources of some important plant polysaccharides for the use as drug delivery excipients are listed in Table 2.1.

References

A.M. Avachat, R.R. Dash, S.N. Shrotriya, Recent investigations of plant based natural gums, mucilages and resins in novel drug delivery systems. Indian J Pharm Educ Res **45**, 86–99 (2011)
C.E. Beneke, A.M. Viljoen, J.H. Hamman, Polymeric plant-derived excipients in drug delivery. Molecules **14**, 2602–2620 (2009)
M.S. Hasnain, A.K. Nayak, R. Singh, F. Ahmad, Emerging trends of natural-based polymeric systems for drug delivery in tissue engineering applications. Sci. J. UBU **1**, 1–13 (2010)
M.S. Hasnain, A.K. Nayak, Chitosan as responsive polymer for drug delivery applications, in *Stimuli Responsive Polymeric Nanocarriers for Drug Delivery Applications, Volume 1, Types and Triggers*, ed. A.S.H. Makhlouf, N.Y. Abu-Thabit (Woodhead Publishing Series in Biomaterials, Elsevier Ltd., 2018b), pp. 581–605
M. Hati, B.K. Jena, S. Kar, A.K. Nayak, Evaluation of anti-inflammatory and anti-pyretic activity of *Carissa carandas* L. leaf extract in rats. J. Pharm. Chem. Biol. Sci. **1**, 18–25 (2014)

B.K. Jena, B. Ratha, S. Kar, S. Mohanta, A.K. Nayak, Antibacterial activity of the ethanol extract of *Ziziphus xylopyrus* Willd. (Rhamnaceae). Int. J. Pharma Res. Rev. **1**, 46–50 (2012a)

B.K. Jena, B. Ratha, S. Kar, S. Mohanta, M. Tripathy, A.K. Nayak, Wound healing potential of *Ziziphus xylopyrus* Willd. (Rhamnaceae) stem bark ethanol extract using *in vitro* and *in vivo* model. J. Drug Deliv. Therapeut. **2**, 41–46 (2012b)

P.B. Malafaya, G.A. Silva, R.L. Reis, Natural-origin polymers as carriers and scaffolds for biomolecules and cell delivery in tissue engineering applications. Adv. Drug Deliv. Rev. **59**, 207–233 (2007)

J.F. Mano, G.A. Silva, H.S. Azevedo, P.B. Malafaya, R.A. Sousa, S.S. Silva, L.F. Boesel, J.M. Oliveira, T.C. Santos, A.P. Marques, N.M. Neves, R.L. Reis, Natural origin biodegradable systems in tissue engineering and regenerative medicine: present status and some moving trends. J. R. Soc. Interf. **4**, 999–1030 (2007)

A.K. Nayak, Advances in therapeutic protein production and delivery. Int. J. Pharm. Pharmaceut. Sci. **2**, 1–5 (2010)

A.K. Nayak, D. Pal (2012) Natural polysaccharides for drug delivery in tissue engineering. Everyman's Sci XLVI, 347–352

A.K. Nayak, D. Pal, K. Santra, Screening of polysaccharides from tamarind, fenugreek and jackfruit seeds as pharmaceutical excipients. Int. J. Biol. Macromol. **79**, 756–760 (2015)

A.K. Nayak, D. Pal, Natural starches-blended ionotropically-gelled micrparticles/beads for sustained drug release, in *Handbook of composites from renewable materials*, ed. V.K. Thakur, M.K. Thakur, M.R. Kessler, Volume 8, Nanocomposites: Advanced Applications (Wiley-Scrivener, USA, 2017a), pp. 527–560

A.K. Nayak, D. Pal, Tamarind seed polysaccharide: an emerging excipient for pharmaceutical use. Indian J. Pharm. Educ. Res. 51, S136-S146 (2017b)

A.K. Nayak, D. Pal, D.R. Pany, B. Mohanty, Evaluation of *Spinacia oleracea* L. leaves mucilage as innovative suspending agent. J. Adv. Pharm. Technol. Res. **1**, 338–341 (2010)

A.K. Nayak, D. Pal, Functionalization of tamarind gum for drug delivery, in: *Functional biopolymers*, ed. V.K. Thakur, M.K. Thakur (Springer International Publishing, Switzerland, 2018), pp. 35–56

A.K. Nayak, H. Bera, M.S. Hasnain, D. Pal, Graft-copolymerization of plant polysaccharides, in: *Biopolymer grafting, synthesis and properties*, ed. V.K. Thakur (Elsevier Inc., 2018a), pp. 1–62

A.K. Nayak, M.S. Hasnain, D. Pal, Gelled microparticles/beads of sterculia gum and tamarind gum for sustained drug release, in *Handbook of springer on polymeric gel*, ed. V.K. Thakur, M.K. Thakur (Springer International Publishing, Switzerland, 2018b), pp. 361–414

D. Pal, A.K. Nayak, S. Kalia, Studies on *Basella alba* L. leaves mucilage: evaluation of suspending properties. Int. J. Drug Discov. Technol. **1**, 15–20 (2010)

D. Pal, A.K. Nayak, Alginates, blends and microspheres: controlled drug delivery, in *Encyclopedia of biomedical polymers and polymeric biomaterials*, ed. M. Mishra (Taylor & Francis Group, New York, NY 10017, U.S.A., Vol. I, 2015a), pp. 89–98

D. Pal, A.K. Nayak, Interpenetrating polymer networks (IPNs): natural polymeric blends for drug delivery, in: *Encyclopedia of biomedical polymers and polymeric biomaterials*, ed. M. Mishra (Taylor & Francis Group, New York, NY 10017, U.S.A., Vol. VI, 2015b), pp. 4120–4130

D. Pal, A.K. Nayak, Plant polysaccharides-blended ionotropically-gelled alginate multiple-unit systems for sustained drug release, in *Handbook of composites from renewable materials, Volume 6, polymeric composites*, ed. V.K. Thakur, M.K. Thakur, M.R. Kessler (Wiley-Scrivener, USA, 2017), pp. 399–400

S. Sinha Mahapatra, S. Mohanta, A.K. Nayak, Preliminary investigation on angiogenic potential of *Ziziphus oenoplia* M. root ethanolic extract by chorioallantoic membrane model. Sci. Asia **37**, 72–74 (2011)

Chapter 3
Gum Arabic Based Multiple Units for Oral Drug Delivery

3.1 Gum Arabic (GAr)

GAr is a plant-derived natural polysaccharidic gum, which is originally extracted from *Acacia nilotica* tree, belonging to the family, Leguminosae (Ali et al. 2009; Dauqan and Abdullah 2013). Recently, GAr is largely extracted from two acacia species, namely *Acacia seyal* and *Acacia senegal* (Heuzé et al. 2016; Patel and Goyal 2015). These acacia plants are cultivated mostly in Arabia, Senegal, Sudan, Somalia and some in countries of West Asia (Webb 2009; Heuzé et al. 2016). GAr is a nontoxic and biodegradable gum (Avadi et al. 2010). As a pharmaceutical biopolymer, GAr possesses various useful physical properties like high solubility in the aqueous medium, stability in various pH milieu, and good quality of gel-forming ability (Dauqan and Abdullah 2013; Gils et al. 2010; Yael et al. 2006). GAr is a branched polysaccharide of slightly acidic in nature and exists as salt-mixture of calcium, potassium, and magnesium polysaccharide acids having a main chain of $(1 \rightarrow 3)$-β-D-galactopyranosyl units and the side chains having L-arabinofuranosyl, L-rhamnopyranosyl, D-galactopyranosyl and D-glucopyranosyl uronic acid units (Gils et al. 2010; Nayak et al. 2012).

3.2 Use of GAr as Pharmaceutical Excipients

GAr is principally utilized as a thickener, suspending agent, emulsifier, and stabilizer in various food products and thus, is known as food hydrocolloids in the food industry (Mariana et al. 2012; Patel and Goyal 2015; Yadav et al. 2007). It is extensively utilized in the formula of various cosmeceutical and pharmaceutical formulations (Avadi et al. 2010). It is also employed as a suspending agent for different insoluble drug candidates (Dauqan and Abdullah 2013). The uses of GAr as a pharmaceutical excipient in various drug releasing formulations are summarized in Table 3.1.

© The Author(s), under exclusive license to Springer Nature Singapore Pte Ltd. 2019
A. K. Nayak and M. S. Hasnain, *Plant Polysaccharides-Based Multiple-Unit Systems for Oral Drug Delivery*, SpringerBriefs in Applied Sciences and Technology, https://doi.org/10.1007/978-981-10-6784-6_3

Table 3.1 Pharmaceutical applications of GAr in different formulations

Formulations made of GAr	Pharmaceutical applications	References
Ofloxacin tablets	Binders	Mistry et al. (2014)
Fast disintegrating tablets of glipizide	Disintegrating agent	Jha and Chetia (2012)
Film-coated immediate release oral tablet formulation	Film-coating former	Alkarib et al. (2016)
Calcium alginate/GAr beads containing glibenclamide	Encapsulating material and release retardant	Nayak et al. (2012)
Insulin nanoparticles using chitosan and GAr	Encapsulating material and release retardant	Avadi et al. (2010)
Monolithic osmotic tablet system	Osmotic, suspending, and expanding agent	Lu et al. (2003)
Self-gelling primaquine–GAr conjugate	Injectable gelling agent	Nishi and Jayakrishnan (2007)
Isoniazid encapsulated chitosan–GAr particles for sustained release formulation	Encapsulating material and release retardant	Kadare et al. (2014)

3.3   GAr–Alginate Beads of Glibenclamide

Nayak et al. (2012) developed GAr–alginate beads for sustained releasing of glibenclamide via the ionotropic gelation by means of ionotropic cross-linking solutions of
calcium chloride. When aqueous dispersion mixtures contained polymers (GAr and
sodium alginate) and drug (glibenclamide) were dropped into the ionotropic cross-
linking solutions, spherically shaped GAr–alginate beads of glibenclamide were
formed instantly as a result of the electrostatic interaction in between carboxylic acids
with a negative charge occurred on the polysaccharidic backbone of sodium alginate
and calcium ions (positively charged) of the ionotropic cross-linking solutions. Furthermore, calcium ions could compete with magnesium ions and potassium ions
occurred in the polymeric structure of GAr. The impacts of sodium alginate amounts
and GAr amounts in the polymer blend on the encapsulation efficiency of glibenclamide in these beads as well as cumulative in vitro drug releasing after 8 h by these
beads were analyzed for the formulation optimization by employing the central composite design and response surface methodology. The optimized GAr–alginate beads
of glibenclamide exhibited encapsulation efficiency of $86.02 \pm 2.97\%$ glibenclamide
and $35.68 \pm 1.38\%$ cumulative in vitro release of glibenclamide after 7 h. The rise in
the drug encapsulation efficiency and the reduction of encapsulated glibenclamide
releasing, in vitro, were detected as the polymer amounts (sodium alginate amounts

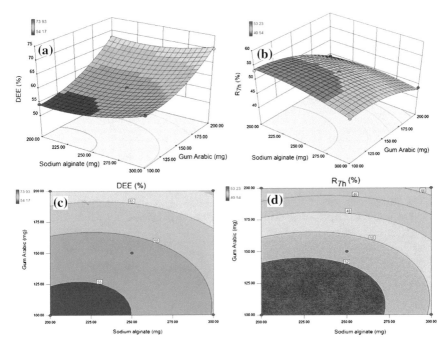

Fig. 3.1 Three-dimensional response surface plots (**a** and **b**) and two-dimensional corresponding contour plots (**c** and **d**) showing the impacts of sodium alginate amounts and GAr amounts in the polymer blend to prepare GAr–alginate beads of glibenclamide on encapsulation efficiency of glibenclamide (DEE %) and cumulative in vitro drug releasing after 7 h (R_{7h} %) [Nayak et al. (2012); Copyright @ 2012, with permission from Elsevier B.V.]

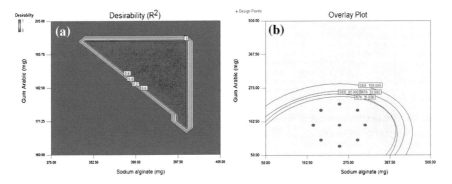

Fig. 3.2 The desirability plot: **a** indicating desirable regression ranges and the overlay plot **b** indicating the region of optimal process variable settings in the formulation optimization of GAr–alginate beads of glibenclamide [Nayak et al. (2012); Copyright @ 2012, with permission from Elsevier B.V.]

(a)   **(b)**

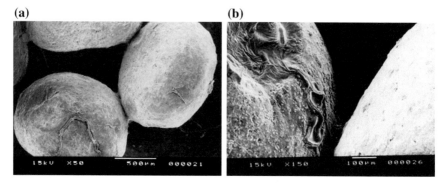

Fig. 3.3 SEM photograph of optimized GAr–alginate beads of glibenclamide (F-O) (**a** and **b**) [Nayak et al. (2012); Copyright @ 2012, with permission from Elsevier B.V.]

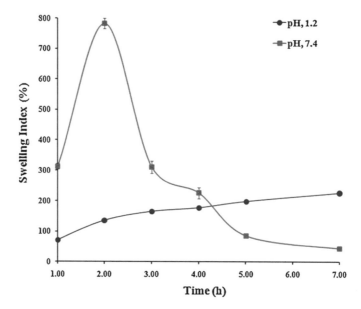

Fig. 3.4 The in vitro swelling of the optimized GAr–alginate beads of glibenclamide in 0.1 N HCl, pH 1.2 and phosphate buffer, pH 7.4 [Nayak et al. (2012); Copyright @ 2012, with permission from Elsevier B.V.]

and GAr amounts) in the polymer blend increased in these GAr–alginate bead formula. Three-dimensional response surface plots and two-dimensional corresponding contour plots showing the impacts of sodium alginate amounts and GAr amounts in the polymer blend to prepare GAr–alginate beads of glibenclamide on encapsulation efficiency of glibenclamide and cumulative in vitro drug releasing after 7 h are presented in Fig. 3.1. The obtained desirability plot indicating desirable regression ranges and the overlay plot indicating the region of optimal process variable settings

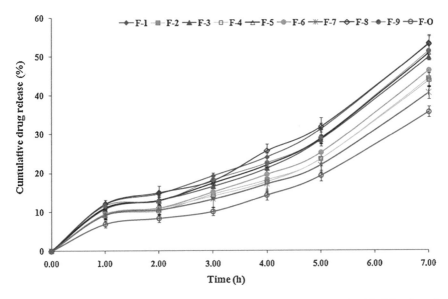

Fig. 3.5 In vitro drug release from various GAr–alginate beads of glibenclamide [Nayak et al. (2012); Copyright @ 2012, with permission from Elsevier B.V.]

are presented in Fig. 3.2. The encapsulation efficiency of glibenclamide in these beads was measured from $54.17 \pm 1.54\%$ to $73.93 \pm 3.37\%$. The bead sizing of these dried GAr–alginate beads of glibenclamide was measured 1.15 ± 0.11 mm to 1.55 ± 0.19 mm. Scanning electron microscopy (SEM) analyses of dried GAr–alginate beads of glibenclamide demonstrated that these beads were spherical in shape with an irregular morphological feature with the rough surfaces. In addition, on the bead surface, typical large wrinkles and cracking occurrences were noticed (Fig. 3.3). The results of Fourier-transform infra red (FTIR) spectroscopy analyses indicated the occurrence of some characteristic peaks of glibenclamide, GAr and sodium alginate and these were emerged in the FTIR pattern devoid of any significant difference or shifting of these characteristic peaks. These results clearly suggested no chemical interaction in between the polymer blends employed (GAr and sodium alginate) and the encapsulated glibenclamide in these GAr–alginate beads of glibenclamide. The in vitro swelling of the optimized GAr–alginate beads of glibenclamide was detected to be controlled by the alteration of pH of the swelling mediums (Fig. 3.4). The in vitro releasing of encapsulated glibenclamide from all these GAr–alginate beads of glibenclamide was evaluated in the acidic milieu (pH 1.2) for initial 2 h and then, in the alkaline milieu (pH 7.4) for the subsequently prolonged period. All these GAr–alginate beads demonstrated an in vitro drug releasing pattern indicating prolonged sustained releasing of encapsulated drug (here glibenclamide) over 7 h (Fig. 3.5). The in vitro drug (glibenclamide) releasing in the acidic milieu (gastric pH) was calculated as <15% after 2 h owing to the shrinkage of ionotropically gelled alginate-based gel beads at the acidic gastric pH. After 2 h of in vitro drug releasing in the acidic milieu (pH 1.2), the in vitro drug releasing was found more rapid in the alkaline milieu (pH 7.4) as because of the high extent of swelling of ionotropically gelled alginate-based gel beads in the alkaline (intestinal pH). This in vitro glibenclamide releasing was calculated to obey the first-order kinetic model with the mechanism of anomalous (non-Fickian) diffusion over 7 h.

References

B.H. Ali, A. Ziada, G. Blunden, Biological effects of gum arabic: a review of some recent research. Food Chem. Toxicol. **47**, 1–8 (2009)

S.Y. Alkarib, D.E. Mohamed Elhassan, A.O. Nur, Evaluation of gum Arabic as a film coating former for immediate release oral tablet formulation. World J. Pharm. Pharmaceut. Sci. **5**, 32–41 (2016)

M.R. Avadi, A.M.M. Sadeghi, N. Mohammadpour, S. Abedin, F. Atyabi, R. Dinarvand, M. Rafiee-Tehrani (2010) Preparation and characterization of insulin nanoparticles using chitosan and Arabic gum with ionotropic gelation method. Nanomed: Nanotechnol. Biol. Med. 6, 58–63

E. Dauqan, A. Abdullah, Utilization of gum Arabic for industries and human health. Am. J. Appl. Sci. **10**, 1270–1279 (2013)

P.S. Gils, D. Ray, P.K. Sahoo, Designing of silver nanoparticles in gum arabic based semi-IPN hydrogel. Int. J. Biol. Macromol. **46**, 237–244 (2010)

V. Heuzé, H. Thiollet, G. Tran, P. Hassoun, D. Bastianelli, F. Lebas (2016) Gum arabic tree (*Acacia senegal*). Feedipedia, a programme by INRA, CIRAD, AFZ and FAO. https://www.feedipedia. org/node/342

A.K. Jha, D. Chetia, Development of natural gum based fast disintegrating tablets of glipizide. Asian J. Pharm. **6**, 282–288 (2012)

P. Kadare, P. Maposa, A. Dube, C.C. Maponga (2014) Encapsulation of isoniazid in chitosan-gum arabic and poly (lactic-co-glycolic acid) PVA particles to provide a sustained release formulation. J. Pharmaceut. Pharmacol. S(1), 1–6

E.X. Lu, Z.Q. Jiang, Q.Z. Zhang, X.G. Jiang, A water-insoluble drug monolithic osmotic tablet system utilizing gum arabic as an osmotic, suspending, and expanding agent. J. Control Rel. **92**, 375–382 (2003)

A.M. Mariana, L.B. María, V. Lorena, D.B. Claudio, Gum Arabic: More than an edible emulsifier. Products Appl. Biopolym. (2012). https://doi.org/10.5772/33783

A.K. Mistry, C.D. Nagda, D.C. Nagda, B.C. Dixit, R.B. Dixit, Formulation and *in vitro* evaluation of ofloxacin tablets using natural gums as binders. Sci. Pharm. **82**, 441–448 (2014)

A.K. Nayak, B. Das, R. Maji, Calcium alginate/gum Arabic beads containing glibenclamide: Development and *in vitro* characterization. Int. J. Biol. Macromol. **51**, 1070–1078 (2012)

K.K. Nishi, A. Jayakrishnan, Self-gelling primaquine-gum arabic conjugate: an injectable controlled delivery system for primaquine. Biomacromol **8**, 84–90 (2007)

S. Patel, A. Goyal, Applications of natural polymer gum Arabic: a review. Int. J. Food Prop. **18**, 986–998 (2015)

J.L.A. Webb, The trade in gum Arabic: prelude to French conquest in Senegal. J. Afr. Hist. **26**, 149–168 (2009)

M.P. Yadav, J.M. Igartuburu, Y. Yan, E.A. Nothnagel, Chemical investigation of the structural basis of the emulsifying activity of gum arabic. Food Hydrocol **21**, 297–308 (2007)

D. Yael, C. Yachin, Y. Rachel, Structure of gum Arabic in aqueous solution. J. Polym.Sci **44**, 3265–3271 (2006)

Chapter 4
Tamarind Polysaccharide Based Multiple Units for Oral Drug Delivery

4.1 Tamarind Polysaccharide (TP)

Tamarind tree (*Tamarindus indica*, family: Fabaceae) is vernacularly well known as *imli* (Hindi) tree (Joseph et al. 2012; Nayak 2010). This evergreen tree is grown in almost all over India and in other Southeast Asian countries as well (Gupta et al. 2010; Samal and Dangi 2014). In India, tamarind extraction is about 0.30 million tons, annually (Nayak 2010; Nayak and Pal 2017). Tamarind seed mainly consists of 67.10 gm/kg of crude fiber with significantly higher carbohydrate content in the sugar form (Nayak 2010). Polysaccharide extracted from tamarind kernel powder (i.e., TP) is a cell wall storage substance (Gupta et al. 2010). TP is still one of the promising economic plant polysaccharides (Avachat et al. 2011; Manchanda et al. 2014). TP was first extracted by Rao et al. (1946) in the research laboratory. The extraction methodology of TP was modified by the same group (Rao and Srivastava 1973), and then, it was further modified on a laboratory scale (Nandi 1975). In the previously reported literature, many researchers have disclosed different methodologies of TP extraction from tamarind kernel powder (Khanna et al. 1987; Samal and Dangi 2014; Tattiyakul et al. 2010). Generally, these methodologies can be categorized into two categories: chemical methodologies and enzymatic methodologies. In the chemical methodology, the powder of tamarind kernel is soaked in the boiling water. The extracted mucilaginous polysaccharide content is then filtered out. The filtered mucilaginous polysaccharide content is introduced to the same volume of acetone to precipitate TP, which is concentrated and then, dried (Gupta et al. 2010). Tamarind kernel powder is mixed up with ethyl alcohol in the enzymatic methodology and then, it is treated with protease (an enzyme) (Tattiyakul et al. 2010). Subsequently, it is centrifuged and then the supernatant is introduced to ethyl alcohol to precipitate TP. The obtained precipitate of TP is collected and then dried.

© The Author(s), under exclusive license to Springer Nature Singapore Pte Ltd. 2019
A. K. Nayak and M. S. Hasnain, *Plant Polysaccharides-Based Multiple-Unit Systems for Oral Drug Delivery*, SpringerBriefs in Applied Sciences and Technology,
https://doi.org/10.1007/978-981-10-6784-6_4

Chemically, TP comprises of $(1 \rightarrow 4)$-β-D-glucan backbone that is substituted with the side chains of α-D-xylopyranose and β-D-galactopyranosyl $(1 \rightarrow 2)$-α-D-xylopyranose linked $(1 \rightarrow 6)$ to glucose residues, where the ratio of glucose, xylose, and galactose units is 2.8:2.25:1.0 (Nayak 2010; Nayak and Pal 2017). TP is known as a galactoxyloglucan. Around, 80% of glucose residues are substituted by the $1 \rightarrow 6$ linked xylose residues and partly substituted by the $1 \rightarrow 2$ galactose residues (Kaur et al. 2012; Lang et al. 1992). The molecular weight range from 2.50×10^5 to 6.50×10^5 is reported for TP (Zhang et al. 2008). TP is a water-soluble and acid-stable polysaccharide (Sumathi and Ray 2002). The higher level of glucan chain substitution in the TP molecules yields a rigid expanded molecular conformation with large volume occupancy in the aqueous milieu (Gupta et al. 2010; Joseph et al. 2012). When dispersed in the aqueous solvents, TP is capable of showing a tendency to be self-aggregated. The aggregates are lateral assemblies of single polysaccharidic strands with a behavior that can be explained well by Kuhn model or the worm-like chain model (Gupta et al. 2010). TP swells in the aqueous medium to form mucilaginous solutions having characteristic features of the non-Newtonian rheology as well as the pseudoplastic behavior (Joseph et al. 2012; Kaur et al. 2012). TP is also described as gel-producing biopolymer having bioadhesivity and mucomimetic characteristics (Kaur et al. 2012; Sahoo et al. 2010). In cold water and in organic solvents like ethanol, methanol, ether, acetone, etc., it is insoluble (Gupta et al. 2010; Nayak 2010). TP is reported as biocompatible, noncarcinogenic, and haemostatic (Avachat et al. 2011). Hepatoprotective, antidiabetic, and anti-inflammatory actions of TP have also been demonstrated in a research by Samal and Dangi (2014). TP has also film forming capability with a high degree of flexibility and good tensile strength (Gupta et al. 2010).

4.2 Use of TP as Pharmaceutical Excipients

TP is one of the emerging most plant-derived biopolysaccharides, which has been exploited as useful excipients in various potential applications in the fields of food, cosmetic and pharmaceutical technologies (Nayak 2010; Nayak and Pal 2017). During the past few decades, TP is being used as potential pharmaceutical excipients like thickener, emulsifier, suspending agent, gelling agent, biomucoadhesive agent, binder, matrix-former and release retardant in various dosage forms like suspensions, emulsions, gels, ophthalmic preparations, nanoparticles, microparticles, spheroids, beads, tablets, buccal and transdermal patches, etc (Avachat et al. 2011; Gupta et al. 2010; Joseph et al. 2012; Nayak 2010). TP is also used as carrier materials to formulate various controlled drug-releasing systems for numerous drug candidates (Chanda et al. 2008; Chandramouli et al. 2012; Kulkarni et al. 2005). It is also used in the development of oral (Nayak and Pal 2017; Nayak et al. 2014, 2016, 2018; Pal and Nayak 2012), ocular (Gheraldi et al. 2000; Mehra et al. 2010), buccal (Avachat et al. 2013; Bangle et al. 2011), colon (Mishra and Khandare 2011), and nasal (Datta and Bandyopadhyay 2006) drug delivery formulations. TP is also used in the formulations of different mucoadhesive drug-releasing systems because of its hydrophilicity and biomucoadhesivity (Bangle et al. 2011; Nayak et al. 2016; Pal and Nayak 2012). The uses of TP as a pharmaceutical excipient in various drug-releasing formulations are summarized in Table 4.1.

4.3   TP Spheroids of Diclofenac Sodium

TP was employed as a drug-release modifier to prepare spheroids containing diclofenac sodium (Kulkarni et al. 2005). These TP spheroids of diclofenac sodium were prepared using the extrusion-spheronization method. The preparation methodology was studied to determine the influences of different process variables on particle shape, size, and size. TP spheroids of diclofenac sodium showed controlled (zero order) releasing, in vitro. The drug-releasing kinetics obeyed Higuchi and Peppas models. The comparative bioavailability evaluation revealed that these TP spheroids were capable of maintaining the releasing of diclofenac sodium for more than 8 h and could improve the absorption as well as the bioavailability of encapsulated diclofenac sodium.

4.4   TP–Alginate Beads of Diclofenac Sodium

Nayak and Pal (2011) developed TP–alginate beads of diclofenac sodium via the ionotropic gelation by means of ionotropic cross-linking solutions of calcium chloride. When dispersion mixtures contained polymers (TP and sodium alginate) and drug (diclofenac sodium) were dropped into the ionotropic cross-linking solutions, spherically shaped TP–alginate beads of diclofenac sodium were formed instantly as a result of the electrostatic interaction in-between carboxylic acids with a negative charge occurred on the polysaccharidic backbone of sodium alginate and calcium ions (positively charged) of the ionotropic cross-linking solutions. In order to examine the impacts of independent process variables such as sodium alginate to TP ratio and cross-linker concentration on the drug encapsulation and releasing of TP–alginate beads of diclofenac sodium, a 3^2-factorial design-assisted optimization technique was employed. The optimization process revealed a significant rise in the efficiency of drug encapsulations and a significant reduction in cumulative in vitro releasing of diclofenac sodium after 10 h with a reduction in the sodium alginate to TP ratio and an increase in $CaCl_2$ concentrations. Figures 4.1 and 4.2 present the impact of sodium alginate to TP ratio and cross-linker concentration on drug encapsulation efficiency and cumulative in vitro releasing of diclofenac sodium after 10 h by the response surface plots and the contour plots, respectively. The optimized TP–alginate beads contained diclofenac sodium clearly showed a bead sizing of 0.71 ± 0.03 mm, encapsulation efficiency of $97.32 \pm 4.03\%$ diclofenac sodium and $69.08 \pm 2.36\%$ of cumulative diclofenac sodium releasing after 10 h. The surface morphological feature

Table 4.1 Pharmaceutical applications of TP in different formulations

Formulations made of TP	Pharmaceutical applications	References
Castor oil emulsions	Emulsifier	Kumar et al. (2011)
Nimesulide suspension	Suspending agent	Deveswaran et al. (2010)
Paracetamol suspension	Suspending agent	Rishabha et al. (2010)
Caffeine tablets	Release retardant	Sumathi and Ray (2002)
Mucoadhesive tablets of tarbutaline sulfate	Mucoadhesive, release retardant	Chanda et al. (2008)
Tramadol HCl tablets	Binder	Phani Kumar et al. (2011)
Ibuprofen tablets	Release retardant	Kulkarni et al. (2011)
Acyclovir matrix tablets	Matrix former, release retardant	Chandramouli et al. (2012)
Aceclofenac matrix tablets	Matrix former, release retardant	Basavaraj et al. (2011)
Verapamil HCl tablets	Matrix former, release retardant	Kulkarni et al. (1997)
Paclitaxel composites	Release retardant	Sahoo et al. (2010)
Diclofenac sodium spheroids	Release retardant	Kulkarni et al. (2005)
Nitrendipine buccal tablet	Mucoadhesive, release retardant	Bangle et al. (2011)
Nifedipine buccoadhesive tablet	Mucoadhesive, release retardant	Patel et al. (2009)
Mucoadhesive buccal films of rizatriptan benzoate	Mucoadhesive, film former	Avachat et al. (2013)
Nasal gel of diazepam	Mucoadhesive, gelling agent	Datta and Bandyopadhyay (2006)
Ibuprofen matrix tablets	Release retardant, colon specific	Mishra and Khandare (2011)
Pilocarpine ocular gels (in situ)	Gelling agent	Mehra et al. (2010)
Gentamicin and ofloxacin ocular gels	Mucoadhesive, gelling agent	Gheraldi et al. (2000)
Microbeads of diclofenac sodium	Matrix-former, release retardant, encapsulating agent	Kulkarni et al. (2012)
Microparticles of aceclofenac	Matrix-former, release retardant, encapsulating agent	Jana et al. (2013)
Mucoadhesive beads of metformin HCl	Matrix-former, release retardant, encapsulating agent, mucoadhesive	Nayak et al. (2014, 2016)
Beads of diclofenac sodium	Matrix-former, release retardant, encapsulating agent	Nayak and Pal (2011)
Mucoadhesive microspheres of gliclazide	Matrix-former, release retardant, encapsulating agent, mucoadhesive	Pal and Nayak (2012)

(a)   **(b)**

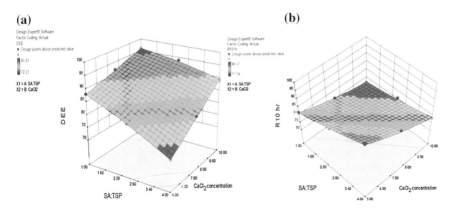

Fig. 4.1 Effect of individual variables (sodium alginate to TP ratio and cross-linker concentration) on drug encapsulation efficiency (DEE, %) and cumulative in vitro releasing of diclofenac sodium after 10 h (R10 h, %) presented by response surface plots (**a** and **b**) [Nayak and Pal (2011); Copyright @ 2011, with permission from Elsevier B.V.]

of optimized TP–alginate beads of diclofenac sodium has already been visualized by scanning electron microscopy (SEM), demonstrating a denser morphological feature with the rough surfaces with absence of cracking occurrences (Fig. 4.3). The nuclear magnetic resonance (^{1}H NMR) and Fourier-transform infrared (FTIR) spectroscopy studies revealed that the encapsulated drug was found compatible with TP and sodium alginate utilized to prepare TP–alginate beads of diclofenac sodium. The in vitro release of encapsulated drug from TP–alginate beads showed a continued release of drugs over 10 h (Fig. 4.4). The release of in vitro drugs was followed by a controlled releasing pattern (zero-order kinetics) with a case-II transport mechanism. The swelling of TP–alginate beads contained diclofenac sodium was lower in 0.1 N HCl than in the phosphate buffer (pH 7.4), initially due to the shrinking of alginate-based hydrogels at the acidic pH environment. At 2–3 h of in vitro swelling in the phosphate buffer (pH 7.4), the utmost swelling pattern of optimized TP–alginate beads of diclofenac sodium was observed, where the erosion and dissolution of the TP–alginate bead-matrix occurred (Figs. 4.5 and 4.6).

4.5 TP–Alginate Mucoadhesive Beads of Metformin HCl

Metformin HCl-encapsulated ionotropically gelled TP–alginate mucoadhesive beads were formulated using calcium chloride solutions as an ionotropic cross-linking solution and evaluated for sustained releasing of encapsulated metformin HCl over a longer period (Nayak and Pal 2013; Nayak et al. 2016). In order to examine the impacts of independent process variables such as sodium alginate to TP ratio and cross-linker concentration on the drug encapsulation and releasing of metformin HCl

(a)   **(b)**

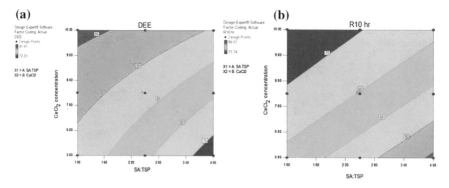

Fig. 4.2 Effect of individual variables (sodium alginate to TP ratio and cross-linker concentration) on drug encapsulation efficiency (DEE, %) and cumulative in vitro releasing of diclofenac sodium after 10 h (R10 h, %) presented by contour plots (**a** and **b**) [Nayak and Pal (2011); Copyright @ 2011, with permission from Elsevier B.V.]

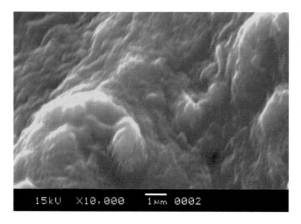

Fig. 4.3 The SEM photograph of the surface of optimized TP–alginate beads of diclofenac sodium [Nayak and Pal (2011); Copyright @ 2011, with permission from Elsevier B.V.]

at 2 h and 10 h, a 3^2-factorial design-assisted optimization technique was employed to develop TP–alginate mucoadhesive beads of metformin HCl. The optimization process revealed a significant rise in the efficiency of metformin HCl encapsulations with a reduction in the sodium alginate to TP ratio and an increase in calcium chloride concentrations in the ionotropic cross-linking solutions. This also revealed a significant reduction in cumulative in vitro releasing of encapsulated metformin HCl after 2 h with the increment of both sodium alginate to TP ratio and ionotropic cross-linker concentration; whereas, a significant reduction in the cumulative metformin HCl release from these beads at 10 h was detected with a reduction in the sodium alginate to TP ratio and an increase of ionotropic cross-linker concentration in the ionotropic cross-linking solutions. All these formulated TP–alginate beads contained metformin HCl clearly showed a bead sizing of 1.05 ± 0.03 mm to 1.52 ±

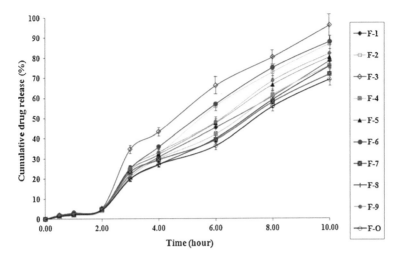

Fig. 4.4 In vitro release of encapsulated drug from TP–alginate beads [Nayak and Pal (2011); Copyright @ 2011, with permission from Elsevier B.V.]

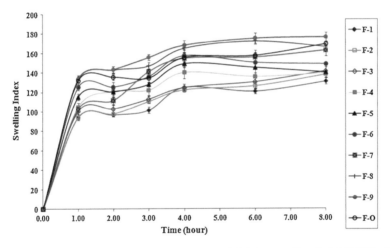

Fig. 4.5 Swelling behavior profile TP–alginate beads of diclofenac sodium in 0.1 N HCl, pH 1.2 [Nayak and Pal (2011); Copyright @ 2011, with permission from Elsevier B.V.]

0.09 mm average diameter and encapsulation efficiency of $71.86 \pm 1.88\%$ to $94.86 \pm 3.92\%$ metformin HCl. The optimized TP–alginate beads contained metformin HCl which displayed a metformin HCl encapsulation efficiency of $94.86 \pm 3.92\%$ and bead sizing average diameter of 1.24 ± 0.07 mm. The surface morphological feature of optimized TP–alginate beads containing metformin HCl was analyzed by SEM, which demonstrated very rough surface morphology with typical larger wrinkles and derbies (Fig. 4.7) (Nayak et al. 2016). FTIR spectra studies revealed that there were no chemical interaction in-between metformin HCl (i.e., encapsulated

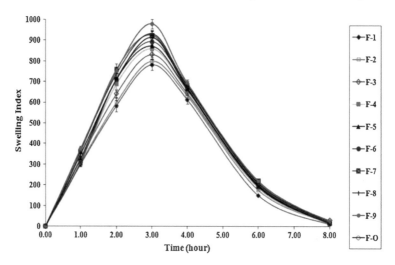

Fig. 4.6 Swelling behavior profile TP–alginate beads of diclofenac sodium in phosphate buffer, pH 7.4 [Nayak and Pal (2011); Copyright @ 2011, with permission from Elsevier B.V.]

drug) and sodium alginate–TP (polymer blends) used to prepare TP–alginate beads of metformin HCl. The in vitro metformin HCl releasing from various TP–alginate beads of metformin HCl was assessed in the acidic milieu (pH 1.2) for initial 2 h and afterward, in the alkaline milieu (pH 7.4) for the subsequently prolonged period. The in vitro metformin HCl releasing in the acidic milieu (pH 1.2) for the initial 2 h was detected slower (<17% of cumulative drug release) after 2 h. On the whole, in vitro metformin HCl releasing results presented a sustained drug releasing pattern over 10 h (Fig. 4.8). A controlled releasing pattern (zero-order kinetics) with a case-II transport mechanism controlled by the swelling was followed by these metformin HCl loaded of TP–alginate beads. The swelling of TP–alginate beads of metformin HCl was lower in the acidic milieu (pH 1.2) than in the alkaline milieu (pH 7.4), initially due to shrinking of alginate-based hydrogels at the acidic pH environment (Fig. 4.9). Ex vivo wash-off behavior of the optimized TP–alginate beads containing metformin HCl onto the biological membrane was detected faster in the intestinal pH (alkaline milieu) as compared to that in the gastric pH (acidic milieu). The overall results of the ex vivo wash-off clearly demonstrated that these optimized TP–alginate beads contained metformin HCl had a high-quality biomucoadhesivity. In vivo performance of the optimized metformin HCl-encapsulated ionotropically gelled TP–alginate mucoadhesive beads was tested using the alloxan-induced diabetic albino rats and the results of this study indicated a significant hypoglycemic outcome over a prolonged period after the oral administration of optimized TP–alginate mucoadhesive beads of metformin HCl.

Fig. 4.7 SEM photograph of optimized TP–alginate beads contained metformin HCl [Nayak et al. (2016); Copyright @ 2015, with permission from Elsevier B.V.]

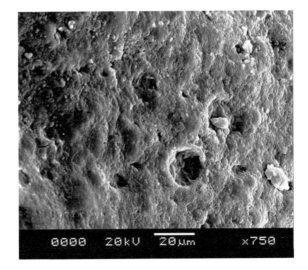

Fig. 4.8 In vitro drug release from optimized TP–alginate beads contained metformin HCl in 0.1 N HCl (pH 1.2) for first 2 h and then, in phosphate buffer (pH 7.4) for the next 8 h [Nayak et al. (2016); Copyright @ 2015, with permission from Elsevier B.V.]

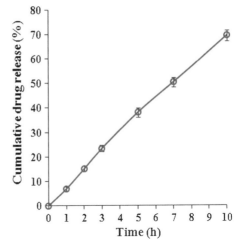

4.6 TP–Alginate Mucoadhesive Beads of Gliclazide

In an investigation, Pal and Nayak (2012) formulated and evaluated of TP–alginate mucoadhesive microspheres for oral administration of gliclazide. These TP-microspheres of gliclazide were prepared by employing ionotropic-gelation method with altering different ratios (1:1 and 1:2) of TP–sodium alginate blends and 2–10% calcium chloride solutions as an ionotropic cross-linking solution. The gliclazide entrapment efficiencies and particle sizes of these TP–alginate microspheres were from $58.12 \pm 2.42\%$ to $82.78 \pm 3.43\%$ and from 752.12 ± 6.42 μm to 948.49 ± 20.92 μm, respectively. The surface morphological feature of TP–alginate micro-

Fig. 4.9 The equilibrium
swelling of optimized
TP–alginate beads contained
metformin HCl in acidic pH
(0.1 N HCl, pH 1.2) and in
alkaline pH (phosphate
buffer, pH 7.4) [Nayak et al.
(2016); Copyright @ 2015,
with permission from
Elsevier B.V.]

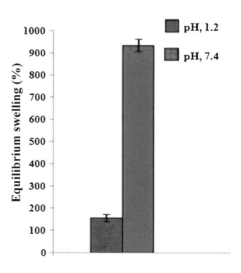

spheres contained gliclazide was analyzed by SEM, which demonstrated spherical
sizes with rough surface morphology. FTIR spectra studies of TP–alginate microspheres contained gliclazide revealed the compatibility of encapsulated gliclazide
within the polymer-blend matrix TP–alginate utilized. In vitro releasing of gliclazide
from these TP–alginate microspheres was tested in the acidic milieu of pH 1.2 and
in the alkaline milieu of pH 7.4. All these TP–alginate microspheres contained gliclazide prepared by employing ionotropic-gelation method revealed a prolonged sustained releasing of gliclazide over 12 h. The sustained gliclazide releasing, in vitro,
from TP–alginate microspheres contained gliclazide was detected to be dependent
on the composition of used polymeric matrix (made of TP–alginate) as well as
the ionotropic cross-linking extent by the divalent calcium ions contained in the
ionotropic cross-linking solutions. The in vitro gliclazide releasing at the acidic
milieu of pH 1.2 was measured relatively slower than that in the alkaline milieu of
pH 7.4. The in vitro swelling performance of these ionotropically gelled TP–alginate
microspheres containing gliclazide was detected to be lower in the acidic milieu of
pH 1.2 as compared to that in the alkaline milieu of pH 7.4. In the acidic milieu, from
$56.50 \pm 4.25\%$ to $70.85 \pm 5.25\%$ of TP–alginate microspheres contained gliclazide
were detected to adhere onto the goat intestinal mucosal membrane; at the same
time as, the adherence of TP–alginate microspheres were found to be varied $38.55
\pm 0.58\%$ to $48.60 \pm 3.25\%$ in the alkaline milieu. The wash-off behavior (ex vivo)
of TP–alginate mucoadhesive microspheres contained gliclazide was detected more
rapidly at the intestinal pH of 7.4 in comparison with that at the gastric pH of 1.2.
The in vivo investigation in the diabetic albino rats (alloxan-induced) revealed that
the best formula of ionotropically gelled TP–alginate mucoadhesive microspheres
contained gliclazide was capable of producing a significant hypoglycaemic potential
after the oral administration of these mucoadhesive microspheres.

4.7 Alginate-Gel-Coated Oil-Entrapped Alginate–TP–Magnesium Stearate Floating Beads of Risperidone

Bera et al. (2015) developed a novel kind of alginate-gel-coated olive oil-entrapped alginate–TP–magnesium stearate floating (buoyant) beads for intragastric delivery of risperidone. The olive oil-entrapped TP–alginate core beads contained magnesium stearate and the magnesium stearate was used here as a low-density substance to impart floatation (buoyancy). The ionotropic gelation mechanism was exploited to prepare these gastroretentive floating beads containing risperidone. The impacts of the ratio of sodium alginate to TP as polymer-blend ratio and the calcium chloride concentration in the ionotropic cross-linker solutions on the drug entrapment efficiency as well as cumulative in vitro drug releasing after 8 h of these floating beads were analyzed for the formulation optimization by employing a 3^2 factorial design-based statistical optimization process. Three-dimensional response surface plots and two-dimensional contour plots describing the effects of the ratio of sodium alginate and TP and calcium chloride concentration on drug encapsulation efficiency and cumulative in vitro drug releasing after 8 h are presented in Fig. 4.10. The uncoated and coated olive-oil-entrapped alginate–TP–magnesium stearate floating beads of risperidone were also characterized by various instrumental analyses such as SEM, FTIR, and powder X-ray diffraction (P-XRD). Both these uncoated and coated olive-oil-entrapped alginate–TP–magnesium stearate floating beads of risperidone showed a spherical morphological structure. Under SEM, the outer surface of the uncoated beads showed rough and fibrous possessing small pores evenly distributed over the entire matrices (Fig. 4.11a, b). On the other hand, alginate-gel-coated beads showed a rough and compact outer surface morphology having pores (Fig. 4.11c, d). The cross-sectional SEM photograph of the coated alginate-gel-coated beads represented sponge-like porous structural morphology due to oil entrapment (Fig. 4.11e, f). The results of the FTIR analysis disqualified the occurrence of any chemical incompatibility in between the encapsulated drug (risperidone) and the polymers used to prepare uncoated and coated olive-oil-entrapped floating beads containing risperidone. The results of P-XRD analyses implied that excipients and encapsulated drug (risperidone) might have interacted at the molecular level. The optimized oil-entrapped alginate–TP–magnesium stearate floating beads of risperidone demonstrated $75.19 \pm 0.75\%$ of drug entrapment efficiency with $78.04 \pm 0.38\%$ of cumulative in vitro drug releasing after 8 h. The developed alginate-gel-coated optimized olive oil-entrapped alginate–TP–magnesium stearate floating beads of risperidone exhibited better-quality buoyancy (floatation) with sustained in vitro drug-releasing performance in the gastric pH (1.2). The density of all these olive-oil-entrapped beads was found less in comparison with that of the simulated gastric fluid (i.e., 1.004 g/cm^3) and accordingly, these beads exhibited floatation having a floating lag time of <8 min. The in vitro drug-releasing patterns of the risperidone-loaded coated and uncoated floating beads were found best fitted in the Higuchi model kinetics with Fickian mechanism and anomalous diffu-

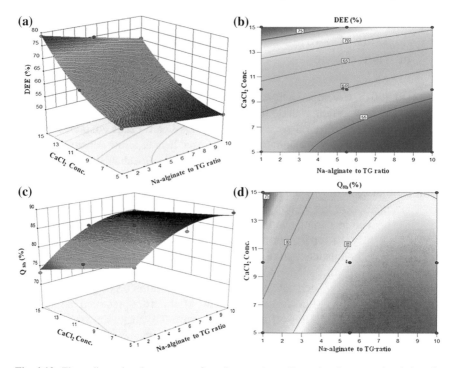

Fig. 4.10 Three-dimensional response surface plots **a** and two-dimensional contour plots **b** describing the effects of ratio of sodium alginate and TP and calcium chloride concentration on drug encapsulation efficiency (DEE %); Three-dimensional response surface plots **c** and two-dimensional contour plots **d** illustrating the effects of ratio of sodium alginate and TP and calcium chloride concentration on cumulative in vitro drug releasing after 8 h (Q8 h%) [Bera et al. (2015); Copyright @ 2015, with permission from Elsevier B.V.]

sion mechanism, respectively. The optimized beads showed a remarkable sustained drug releasing pattern in comparison with the marketed immediate releasing formulation (Fig. 4.12). The swelling profile of the uncoated optimized oil-entrapped alginate–TP–magnesium stearate floating beads of risperidone demonstrated that a lesser extent of in vitro swelling in the acidic pH with a slow augmentation up to 8 h (Fig. 4.13). Therefore, these newly developed alginate-gel-coated oil-entrapped alginate–TP–magnesium-stearate-floating beads of risperidone were found suitable for the use in intragastric delivery of risperidone over a longer time.

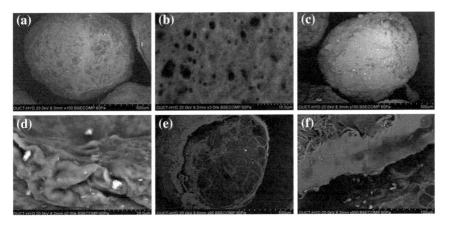

Fig. 4.11   SEM photographs of the uncoated and coated olive-oil-entrapped alginate–TP–magnesium stearate floating beads of risperidone depicting rough and fibrous surface of the uncoated beads (**a**) with small pores (**b**), rough and compact surface of the coated beads (**c**) without pores (**d**); cross-sectional view of the coated beads with less zoom ($95\times$) (**e**) and high zoom ($550\times$) (**f**) [Bera et al. (2015); Copyright @ 2015, with permission from Elsevier B.V.]

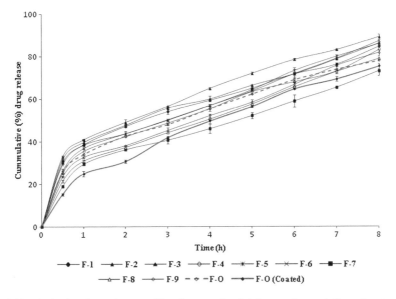

Fig. 4.12   The in vitro drug-release profiles of uncoated and alginate-gel-coated olive-oil-entrapped alginate–TP–magnesium stearate floating beads of risperidone in simulated gastric fluid (pH 1.2) [Bera et al. (2015); Copyright @ 2015, with permission from Elsevier B.V.]

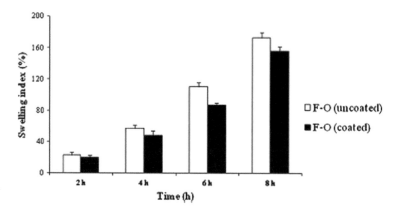

Fig. 4.13 The swelling behavior of the uncoated and alginate-gel-coated-olive oil-entrapped alginate–TP–magnesium stearate floating beads of risperidone in simulated gastric fluid (pH 1.2) [Bera et al. (2015); Copyright @ 2015, with permission from Elsevier B.V.]

4.8 Oil-Entrapped TP–Alginate Floating Beads of Diclofenac Sodium

Nayak et al. (2013) developed groundnut-oil (a low-density oil)-entrapped TP–alginate floating beads for gastroretentive delivery of diclofenac sodium. These floating beads were prepared via ionotropic emulsion gelation using calcium chloride solutions as an ionotropic cross-linking solution. Groundnut oil was made to be entrapped in these beads to achieve the buoyancy (floatation) for a longer duration. When emulsions containing polymers (TP and sodium alginate) and drug (diclofenac sodium) were dropped into the ionotropic cross-linking solutions, spherically shaped gelled groundnut-oil-entrapped TP–alginate floating beads of diclofenac sodium were formed instantly as a result of the electrostatic interaction in between carboxylic acids with a negative charge occurred on the polysaccharidic backbone of sodium alginate and calcium ions (with a positive charge) of the ionotropic cross-linking solutions. In order to examine the impacts of independent process variables such as groundnut oil to water ratio, sodium alginate to TP ratio, and polymer to drug ratio, and on the responses variables such as density of these floating beads, drug entrapment efficiency, and cumulative diclofenac sodium releasing after 8 h in the simulated gastric fluid (pH 1.2), a 2^3-factorial design-assisted optimization technique was employed to develop groundnut oil-entrapped TP–alginate gastroretentive floating beads of diclofenac sodium. The optimized groundnut oil-entrapped TP–alginate floating beads of diclofenac sodium demonstrated 0.88 ± 0.07 g/cm^3 of density and $82.48 \pm 2.34\%$ of drug entrapment efficiency. The results of in vitro buoyancy in simulated gastric fluid (pH 1.2) of these polymer-based floating beads demonstrated excellent buoyancy over 8 h and less than 10 min of floating lag time because of the low density of these beads. The in vitro releasing of diclofenac sodium from these floating beads in the medium of gastric pH (1.2), which clearly demonstrated a sustained drug-releasing pattern over 8 h. This also indicated controlled releasing of encapsulated diclofenac sodium with the release mechanism of super case-II transport. Anti-inflammatory result of the optimized groundnut-oil-entrapped TP–alginate floating beads of diclofenac sodium was analyzed by the rat-paw edema model (carrageenan-induced). After the oral administration of diclofenac sodium (pure) and optimized groundnut-oil-entrapped TP–alginate floating beads of diclofenac sodium, the mean rise in the paw volumes was found to be lower ($p < 0.05$) in comparison with that of the 0.5% w/v sodium carboxymethyl cellulose solution. After the 2 h of oral administration of optimized floating beads, the mean paw volume was found to be lower ($p < 0.05$) in comparison with that of the diclofenac sodium (pure). The optimized groundnut-oil-entrapped TP–alginate floating beads of diclofenac sodium demonstrated a slower inhibition of the paw edema (decrease in paw volumes of the treated rats) in comparison with that of the diclofenac sodium (pure) to maintain the increased inhibition of paw edema over 8 h.

4.9 TP–Alginate Interpenetrated Polymer Network (IPN) Microbeads Containing Diltiazem-Indion 254® Complex

Kulkarni et al. (2012) prepared and characterized IPN microbeads made of TP and sodium alginate for the use in controlled releasing of diltiazem HCl. These TP–alginate IPN microbeads were prepared via the combined ionotropic-gelation/covalent cross-linking approach. In this work, Indion 254® (a cation exchange resin) was used to prepare a drug-resin complex of diltiazem-Indion 254®. Within these TP–alginate IPN microbeads, the diltiazem-Indion 254® complex was encapsulated and these microbeads demonstrated drug entrapment efficiency of 78.15–92.15%. The diltiazem encapsulation efficiency of these TP–alginate IPN microbeads was found to be enhanced with the declining of calcium chloride concentration in the cross-linking solution. At the elevated calcium chloride concentration, more calcium ions was diffused into the drug-resin complex (here diltiazem-Indion 254® complex). As a result, an elevated quantity of diltiazem was moved from the complex. Then, this free drug (diltiazem) was diffused out of the TP–alginate IPN microbeads and this occurrence caused a decrease in the drug entrapment efficiency results. The microbead sizes were measured to be varied within the range of 986–1257 mm. As the concentration of the covalent cross-linker (glutaraldehyde) increased, the size was of TP–alginate IPN microbeads contained diltiazem-Indion 254® complex decreased. This could be because of the arrangement of a stiffer IPN microbead matrix at the elevated cross-linking densities. The results of FTIR analyses demonstrated the configuration of the IPN microbead matrix among TP and alginate within the developed diltiazem-Indion 254® complex encapsulated TP–alginate IPN microbeads contained. In addition, DSC and X-ray diffraction (XRD) analyses demonstrated the physical nature as well as the stability of the drug (i.e., diltiazem) in the IPN microbead matrix made of TP and alginate. The releasing of loaded drug from these TP–alginate-based microbeads formulated with the uncomplexed diltiazem exhibited in vitro drug releasing up to 4 h and those TP–alginate IPN microbeads contained diltiazem-Indion 254® complex demonstrated in vitro drug releasing up to 9 h. The in vivo assessment of these TP–alginate IPN microbeads contained diltiazem-Indion 254® complex was carried out in the Wister rats. The results of the in vivo analyses demonstrated a comparatively elevated plasma drug profile (AUC) and this clearly indicated the better diltiazem bioavailability, in vivo.

4.10 TP-Chitosan IPN Microparticles of Aceclofenac

Jana et al. (2013) prepared IPN microparticles made of TP and chitosan for sustained releasing of aceclofenac. These IPN-based biopolymeric microparticles were prepared via the covalent cross-linking using glutaraldehyde (a covalent cross-linker) at pH 5.5. The entrapment efficiency of aceclofenac in these TP-chitosan IPN microparticles was measured from $85.84 \pm 1.75\%$ to $91.97 \pm 1.30\%$. An increased entrapment efficiency of aceclofenac in these TP-chitosan IPN microparticles was detected as the glutaraldehyde concentration in the cross-linking solution was increased. Thus, the occurrence of increasing entrapment efficiency of aceclofenac in these microparticles could be because of the higher extent of cross-linking by glutaraldehyde. The mean sizes of these IPN microparticles were measured from 490.55 ± 23.24 μm to 621.60 ± 53.57 μm. Reducing the sizes of these IPN microparticles was detected as the glutaraldehyde concentration increased. This occurrence could be by reason of the configuration of stiffer polymeric network by the higher extent of cross-linking density. The microphotographs (SEM images) of aceclofenac-encapsulated TP-chitosan IPN microparticles indicated that these microparticles were of almost spherical in shape (Fig. 4.14). The microparticle surface topographical view was rough with some typical wrinkles that might be occurred as a result of partly collapsing of the gelled network during the drying of formulations. The FTIR spectroscopic results demonstrated the formation of IPN structure in between TP and chitosan. The possible chemical reaction for the formation of IPN structure in between TP and chitosan was also suggested by Jana et al. (2013) (Fig. 4.15). In addition, the results of DSC analyses suggested the stability of encapsulated drug (aceclofenac) in the IPN matrix. In vitro releasing of the drug from these aceclofenac-encapsulated TP-chitosan IPN microparticles was tested by using dialysis bag diffusion in the phosphate buffer, pH 6.8. The obtained results of in vitro aceclofenac releasing demonstrated a sustained release pattern over 8 h (Fig. 4.16). The increasing sustained drug release pattern from these aceclofenac-encapsulated TP-chitosan IPN microparticles was noticed as the glutaraldehyde concentration increased, which might be because of the higher extent of cross-linking by glutaraldehyde. The in vitro releasing of aceclofenac from these TP-chitosan IPN microparticles was found to obey the Korsmeyer–Peppas model and anomalous (non-Fickian) diffusion mechanism over 8 h. In vivo anti-inflammatory action of these aceclofenac-encapsulated TP-chitosan IPN microparticles was ana-

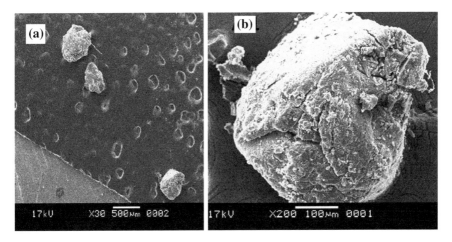

Fig. 4.14 SEM photograph of aceclofenac-encapsulated TP-chitosan IPN microparticles at 30× (**a**) and 200× (**b**) magnifications [Jana et al. (2013); Copyright @ 2013, with permission from Elsevier B.V.]

lyzed by the rat-paw edema model (carrageenan-induced) and the results demonstrated an extended anti-inflammatory action in the carrageenan-induced rats after the oral intake of these aceclofenac-encapsulated TP-chitosan IPN microparticles (Fig. 4.17).

4.11 TP–Gellan Gum Mucoadhesive Beads of Metformin HCl

Nayak et al. (2014) developed TP–gellan gum mucoadhesive beads for oral controlled delivery of metformin HCl via the ionotropic-gelation by means of ionotropic cross-linking solutions of calcium chloride. The formulation optimization of these TP–gellan gum beads of metformin HCl was carried out by means of a 3^2-factorial design-assisted optimization process. In the optimization process, numerical optimization analysis and response surface methodology were employed and analyzed, where the impacts of polymer-blend ratio (gellan gum to TP ratio) and cross-linker concentration on the dependent responses like drug encapsulation efficiency and cumulative drug release after 10 h of TP–gellan gum beads of metformin HCl were optimized. The optimization process revealed a significant rise in the efficiency of metformin HCl encapsulations with a reduction of both gellan gum to TP ratio and calcium chloride concentration. It was also revealed that there was a significant reduction in cumulative in vitro releasing of encapsulated metformin HCl after 10 h with the decreasing of both gellan gum to TP ratio and ionotropic cross-linker concentration. Three-dimensional response surface plots (a and b) and two-dimensional

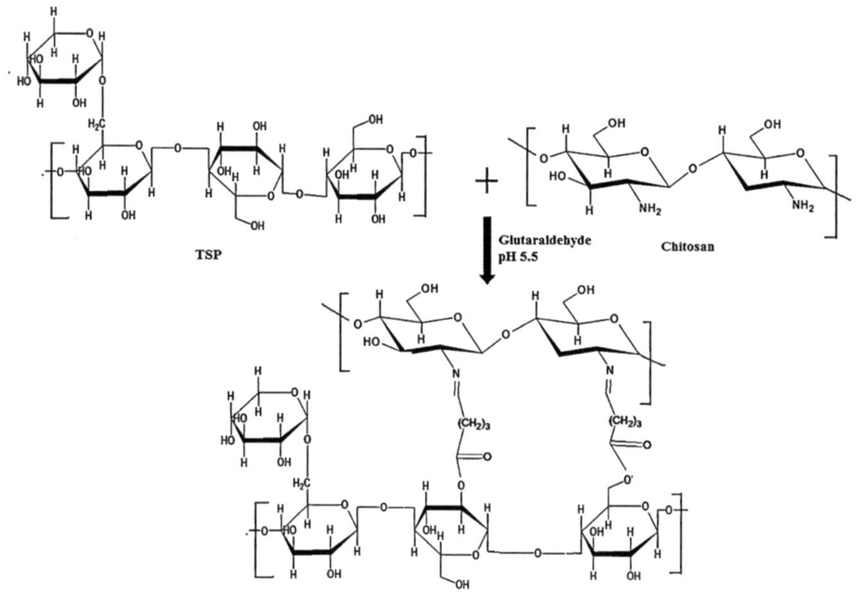

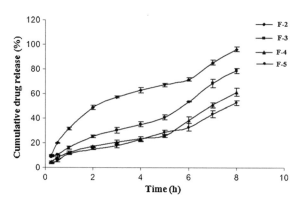

Fig. 4.15 Possible chemical reaction for the formation of IPN structure in between TP and chitosan [Jana et al. (2013); Copyright @ 2013, with permission from Elsevier B.V.]

Fig. 4.16 The in vitro aceclofenac releasing demonstrated a sustained release pattern over 8 h [Jana et al. (2013); Copyright @ 2013, with permission from Elsevier B.V.]

corresponding contour plots (c and d) showing the impacts of gellan gum to TP ratio and cross-linker concentration to prepare TP–gellan gum beads of metformin HCl on encapsulation efficiency of metformin HCl and cumulative drug release after 10 h are presented in Fig. 4.18. The obtained desirability plot indicating desirable regression ranges and the overlay plot indicating the region of optimal process variable settings are presented in Fig. 4.19. The optimized TP–gellan gum beads of metformin HCl clearly exhibited a bead sizing of 1.70 ± 0.24 mm and encapsulation efficiency of $95.73 \pm 4.02\%$ metformin HCl. An increased bead size was measured as both the

Fig. 4.17 The percentages inhibition of paw oedema swelling in Carageenan-induced rat-paw edema model for the standard (pure aceclofenac) and the test (F-5 aceclofenac-encapsulated TP-chitosan IPN microparticles) at various time intervals [Jana et al. (2013); Copyright @ 2013, with permission from Elsevier B.V.]

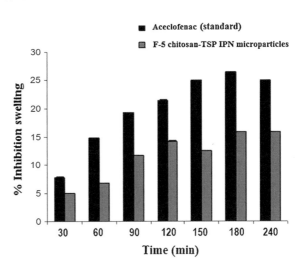

factors (gellan gum to TP and cross-linker concentration) decreased. The surface morphological feature of optimized TP–gellan gum beads of metformin HCl was analyzed by SEM and the SEM results demonstrated that these beads were spherical in shape with an irregular morphological feature with the rough surfaces. In addition, on the bead surface, typical large wrinkles and cracking occurrences were noticed (Fig. 4.20). Furthermore, the presence of little polymer derbies as well as metformin crystals on the TP–gellan gum bead surface was visualized. The in vitro release of encapsulated drug (metformin HCl) from TP–gellan gum beads demonstrated a prolonged drug-releasing pattern over 10 h (Fig. 4.21). The in vitro metformin HCl releasing in the acidic milieu (pH 1.2) for the initial 2 h was detected to be slower (<15.30% of cumulative drug release) after 2 h and subsequently, a more rapid metformin HCl releasing pattern was detected in the alkaline milieu (pH 7.4), comparatively. A controlled releasing pattern (zero-order kinetics) with a case-II transport mechanism controlled by the swelling was followed by these TP–gellan gum beads of metformin HCl. The swelling of TP-gellan beads of metformin HCl was lower in the acidic milieu (pH 1.2) than in the alkaline milieu (pH 7.4), initially due to shrinking of ionotropically gelled gellan gum-based hydrogels at the acidic pH environment (Fig. 4.22a). Ex vivo wash-off behavior of the optimized TP–gellan gum beads of metformin HCl onto the goat intestinal mucosa demonstrated an good-quality biomucoadhesivity, which was detected faster in the intestinal pH (alkaline milieu) as compared to that in the gastric pH (acidic milieu) (Fig. 4.22b). In vivo performance of the optimized TP–gellan gum mucoadhesive beads of metformin HCl was tested using the alloxan-induced diabetic albino rats. The results of this in vivo study indicated a significant hypoglycemic outcome over prolonged period after the oral administration of optimized TP–gellan gum mucoadhesive beads of metformin HCl (Fig. 4.23).

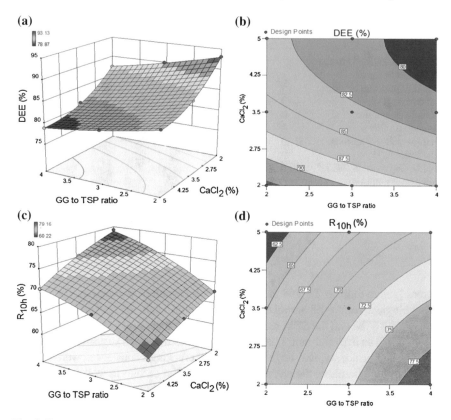

Fig. 4.18 Three-dimensional response surface plots (**a** and **b**) and two-dimensional corresponding contour plots (**c** and **d**) showing the impacts of gellan gum to TP ratio and cross-linker concentration to prepare TP–gellan gum beads of metformin HCl on encapsulation efficiency of metformin HCl (DEE %) and cumulative drug release after 10 h (R_{10h} %) [Nayak et al. (2014); Copyright @ 2013, with permission from Elsevier Ltd.]

4.12   TP-Pectinate Mucoadhesive Beads of Metformin HCl

Nayak et al. (2014) developed mucoadhesive beads of metformin HCl, which was made of TP–low-methoxy pectin polymer blends. In this work, TP-pectinate mucoadhesive beads of metformin HCl were formulated via the ionotropic gelation by means of ionotropic cross-linking solutions of calcium chloride. For the formulation optimization of these biopolymeric beads of metformin HCl, a 3^2-factorial design-assisted optimization process was employed. The impacts of two independable variables like TP amounts and low-methoxy pectin amounts on the dependable responses like encapsulation efficiency of metformin HCl and cumulative metformin HCl releasing after 10 h were examined by means of the response surface

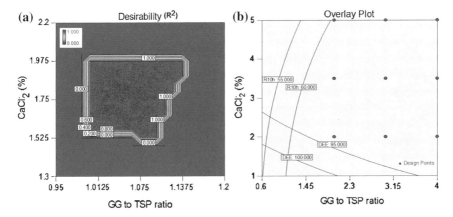

Fig. 4.19 The desirability plot: **a** indicating desirable regression ranges and the overlay plot **b** indicating the region of optimal process variable settings in the formulation optimization of TP–gellan gum beads of metformin HCl [Nayak et al. (2014); Copyright @ 2013, with permission from Elsevier Ltd.]

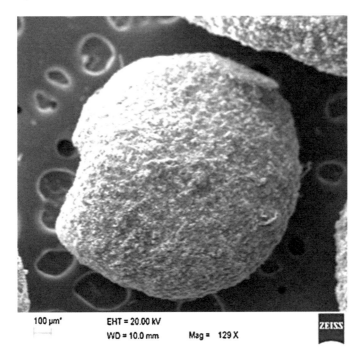

Fig. 4.20 SEM photograph of optimized TP–gellan gum beads of metformin HCl [Nayak et al. (2014); Copyright @ 2013, with permission from Elsevier Ltd.]

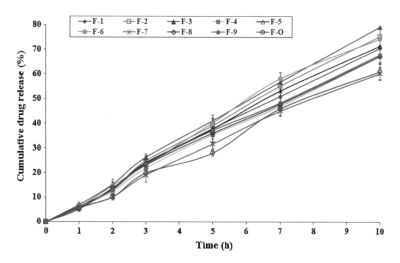

Fig. 4.21 In vitro drug release from various TP–gellan gum beads of metformin HCl [Nayak et al. (2014); Copyright @ 2013, with permission from Elsevier Ltd.]

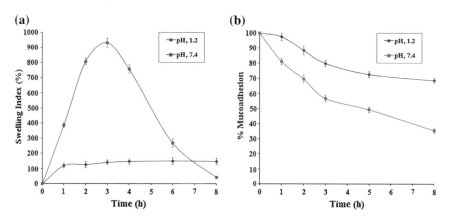

Fig. 4.22 **a** Swelling behavior of optimized TP–gellan gum beads of metformin HCl in 0.1 N HCl (pH 1.2), and phosphate buffer (pH 7.4); **b** Mucoadhesive behavior of optimized TP–gellan gum beads of metformin HCl in 0.1 N HCl (pH 1.2), and phosphate buffer (pH 7.4) [Nayak et al. (2014); Copyright @ 2013, with permission from Elsevier Ltd.]

methodology. The augmentation of encapsulation efficiency of metformin HCl and reduction in the cumulative metformin HCl releasing after 10 h were noticed as both the polymer contents (low-methoxy pectin and TP) increased in the formulated

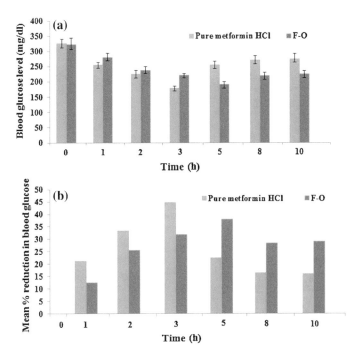

Fig. 4.23 a Comparative in vivo blood glucose level in alloxan-induced diabetic rats after oral administration of pure metformin HCl (standard) and optimized TP–gellan gum mucoadhesive beads of metformin HCl. **b** Comparative in vivo mean percentage reduction in blood glucose level in alloxan-induced diabetic rats after oral administration of pure metformin HCl (standard) and optimized TP–gellan gum mucoadhesive beads of metformin HCl [Nayak et al. (2014); Copyright @ 2013, with permission from Elsevier Ltd.]

TP-pectinate beads. Three-dimensional response surface plots and two-dimensional corresponding contour plots showing the effects of TP amounts and low-methoxy pectin amounts on encapsulation efficiency of metformin HCl and cumulative metformin HCl releasing after 10 h are presented in Fig. 4.24a–d, respectively. The obtained desirability plot indicating desirable regression ranges and the overlay plot indicating the region of optimal process variable settings are presented in Fig. 4.24e, f, respectively. The optimized TP-pectinate mucoadhesive beads of metformin HCl clearly demonstrated a bead sizing of 1.93 ± 0.26 mm and encapsulation efficiency of $95.12 \pm 4.26\%$ metformin HCl. Moreover, the increased bead sizing was measured as the contents of polymers increased in the bead formula and the increment in bead sizing was noticed directly proportional with the rising of encapsulation efficiency of metformin HCl. The surface morphological feature of optimized TP-pectinate beads of metformin HCl was analyzed by SEM. The SEM photograph demonstrated

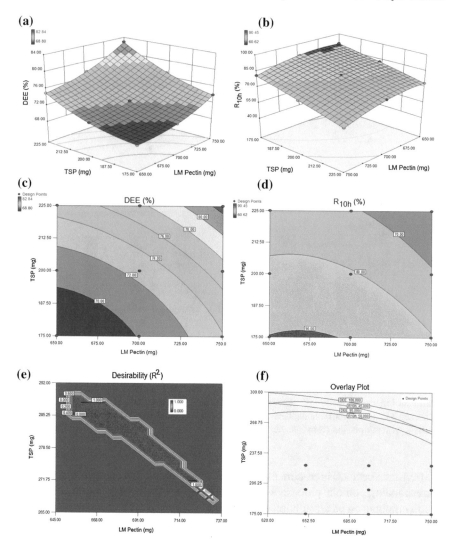

Fig. 4.24 Three-dimensional response surface plots showing the effects of TP amounts and low-methoxy pectin amounts on **a** encapsulation efficiency of metformin HCl (DEE %) and **b** cumulative metformin HCl releasing after 10 h (R_{10h} %); two-dimensional corresponding contour plots showing the effects of TP amounts and low-methoxy pectin amounts on **a** encapsulation efficiency of metformin HCl (DEE %) and **b** cumulative metformin HCl releasing after 10 h (R_{10h} %) the desirability (R^2) plot **e** indicating desirable regression ranges and the overlay plot **f** indicating the region of optimal process variable settings [Nayak et al. (2014); Copyright @ 2013, with permission from Elsevier Ltd.]

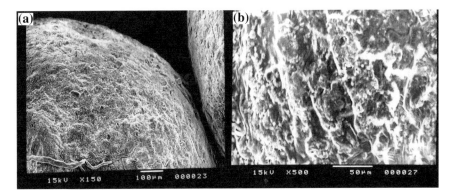

Fig. 4.25 SEM photograph of the optimized TP-pectinate beads of metformin HCl [Nayak et al. (2014); Copyright @ 2013, with permission from Elsevier Ltd.]

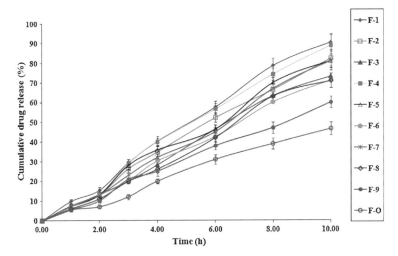

Fig. 4.26 In vitro drug release from various TP-pectinate beads of metformin HCl [Nayak et al. (2014); Copyright @ 2013, with permission from Elsevier Ltd.]

spherically shaped beads having a very rough surface morphology with typical larger wrinkles and polymer derbies (Fig. 4.25). FTIR spectra studies revealed that there were no chemical interaction in between metformin HCl (i.e., encapsulated drug) and low-methoxy pectin–TP (polymer blends) used to prepare TP-pectinate beads of metformin HCl. The in vitro metformin HCl releasing behavior of these TP-pectinate beads demonstrated a prolonged releasing pattern of encapsulated metformin HCl over 10 h (Fig. 4.26). The in vitro metformin HCl releasing from various TP-pectinate

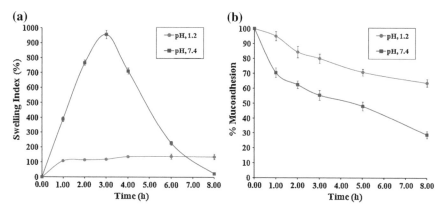

Fig. 4.27 **a** Swelling behavior of optimized TP-pectinate beads of metformin HCl in 0.1 N HCl (pH 1.2), and phosphate buffer (pH 7.4); **b** mucoadhesive behavior of optimized TP-pectinate beads of metformin HCl in 0.1 N HCl (pH 1.2), and phosphate buffer (pH 7.4) [Nayak et al. (2014); Copyright @ 2013, with permission from Elsevier Ltd.]

beads of metformin HCl was assessed in the acidic milieu (pH 1.2) for initial 2 h and afterward, in the alkaline milieu (pH 7.4) for subsequently prolonged period. The in vitro metformin HCl releasing in the acidic milieu (pH 1.2) for initial 2 h was detected to be slower (<15% of cumulative drug release) after 2 h as a result of compactness of the gelled polymeric network by the shrinkage of pectinate gel at acidic milieu (pH 1.2). A controlled releasing pattern (zero-order kinetics) for the encapsulated metformin HCl with a case-II transport mechanism was followed by these TP-pectinate beads of metformin HCl over 10 h. The results of in vitro swelling of optimized TP-pectinate beads of metformin HCl demonstrated a pH-dependent swelling pattern (Fig. 4.27a). In addition, the results of ex vivo wash-off study demonstrated a good mucoadhesive behavior onto the goat intestinal mucosa by the optimized TP-pectinate beads of metformin HCl (Fig. 4.27b). In vivo performance of the optimized TP-pectinate mucoadhesive beads of metformin HCl was tested using the diabetic albino rats. The results of this study indicated a significant hypoglycemic outcome over a prolonged period after the oral administration of optimized TP-pectinate mucoadhesive beads of metformin HCl (Fig. 4.28a, b).

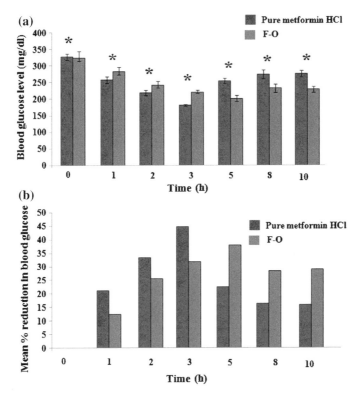

Fig. 4.28 a Comparative in vivo blood glucose level in alloxan-induced diabetic rats after oral administration of pure metformin HCl (standard) and optimized TP-pectinate beads of metformin HCl; **b** Comparative in vivo mean percentage reduction in blood glucose level in alloxan-induced diabetic rats after oral administration of pure metformin HCl (standard) and optimized TP-pectinate beads of metformin HCl [Nayak et al. (2014); Copyright @ 2013, with permission from Elsevier Ltd.]

References

A.M. Avachat, R.R. Dash, S.N. Shrotriya, Recent investigations of plant based natural gums, mucilages and resins in novel drug delivery systems. Indian J. Pharm. Educ. Res. **45**, 86–99 (2011)

A. Avachat, K.N. Gujar, K.V. Wagh, Development and evaluation of tamarind seed glucan-based mucoadhesive buccal films of rizatriptan benzoate. Carbohydr. Polym. **91**, 537–542 (2013)

G.S. Bangle, G.V. Shinde, D.G. Umalkar, K.S. Rajesh, Natural mucoadhesive material based buccal tablets of nitrendipine–formulation and *in-vitro* evaluation. J. Pharm. Res. **4**, 33–38 (2011)

B.V. Basavaraj, B. Someswara Rao, S.V. Kulkarni, P. Patil, C. Surpur, Design and characterization of sustained release aceclofenac matrix tablets containing tamarind seed polysaccharide. Asian J. Pharm. Tech. **1**, 17–21 (2011)

H. Bera, S. Boddupalli, S. Nandikonda, S. Kumar, A.K. Nayak, Alginate gel-coated oil-entrapped alginate–tamarind gum–magnesium stearate buoyant beads of risperidone. Int. J. Biol. Macromol. **78**, 102–111 (2015)

R. Chanda, S.K. Mahapatro, T. Mitra, A. Roy, S. Bahadur, Development of oral mucoadhesive tablets of tarbutalline sulphate using some natural materials extracted from *Albelmoschus esculeatus* and *Tamarindus indica*. Res. J. Pharm. Res. **1**, 46–51 (2008)

Y. Chandramouli, S. Firoz, A. Vikram, C. Padmaja, R.N. Chakravarthi, Design and evaluation of controlled release matrix tablets of acyclovir sodium using tamarind seed polysaccharide. J. Pharm. Biol. **2**, 55–62 (2012)

R. Datta, A.K. Bandyopadhyay, A new nasal drug delivery system for diazepam using natural mucoadhesive polysaccharide obtained from tamarind seeds. Saudi Pharm. J. **14**, 115–119 (2006)

R. Deveswaran, S. Bharath, S. Furtado, S. Abraham, B.V. Basavaraj, V. Madhavan, Isolation and evaluation of tamarind seed polysaccharide as a natural suspending agent. Int. J. Pharm. Biol. Arch. **1**, 360–363 (2010)

E. Gheraldi, A. Tavanti, F. Celandroni, A. Lupetti, C. Blandizzi, E. Boldrini, M. Campa, S. Senesi, Effect of novel mucoadhesive polysaccharide obtained from tamarind seeds on the intraocular penetration of gentamicin and ofloxacin in rabbits. J. Antimicrob. Chemother. **46**, 831–834 (2000)

V. Gupta, R. Puri, S. Gupta, S. Jain, G.K. Rao, Tamarind kernel gum: An upcoming natural polysaccharide. Sys. Rev. Pharm. **1**, 50–55 (2010)

S. Jana, A. Saha, A.K. Nayak, K.K. Sen, S.K. Basu, Aceclofenac-loaded chitosan-tamarind seed polysaccharide interpenetrating polymeric network microparticles. Colloids Surf. B: Biointerf. **105**, 303–309 (2013)

J. Joseph, S.N. Kanchalochana, G. Rajalakshmi, V. Hari, R.D. Durai, Tamarind seed polysaccharide: a promising natural excipients for pharmaceuticals. Int. J. Green Pharm. **6**, 270–278 (2012)

H. Kaur, S. Yadav, M. Ahuja, N. Dilbaghi, Synthesis, characterization and evaluation of thiolated tamarind seed polysaccharide as a mucoadhesive polymer. Carbohydr. Polym. **90**, 1543–1549 (2012)

M. Khanna, R.C. Nandi, J.P. Sarin, Standardization of tamarind seed polyose for pharmaceutical use. Indian Drugs **24**, 268–269 (1987)

D. Kulkarni, A.K. Dwivedi, J.P.S. Sarin, S. Singh, Tamarind seed polyose: a potential polysaccharide for sustained release of verapamil hydrochloride as a model drug. Indian J. Pharm. Sci. **59**, 1–7 (1997)

G.T. Kulkarni, K. Gowthamarajan, R.R. Dhobe, F. Yohanan, B. Suresh, Development of controlled release spheroids using natural polysaccharide as release modifier. Drug Deliv. **12**, 201–206 (2005)

G.T. Kulkarni, P. Seshababu, S.M. Kumar, Effect of tamarind seed polysaccharide on dissolution behaviour of ibuprofen tablets. J. Chronother Drug Deliv. **2**, 49–56 (2011)

R.V. Kulkarni, S. Mutalik, B.S. Mangonda, U.Y. Nayak, Novel interpenetrated polymer network microbeads of natural polysaccharides for modified release of water soluble drug: *in-vitro* and *in-vivo* evaluation. J. Pharm. Pharmacol. **64**, 530–540 (2012)

R. Kumar, S.R. Patil, M.B. Patil, M.S. Paschapur, R. Mahalaxmi, Isolation and evaluation of the emulsifying properties of tamarind seed polysaccharide on castor oil emulsion. Der. Pharm. Lett. **2**, 518–527 (2011)

P. Lang, G. Masci, M. Dentini et al., Tamarind seed polysaccharide: Preparation, characterization and solution properties of carboxymethylated, sulphated and alkyaminated derivatives. Carohydr. Polym. **17**, 185–198 (1992)

R. Manchanda, S.C. Arora, R. Manchanda, Tamarind seed polysaccharide and its modification-versatile pharmaceutical excipients-a review. Int. J. Pharm. Tech. Res. **6**, 412–420 (2014)

G.R. Mehra, M. Manish, S. Rashi, G. Neeraj, D.N. Mishra, Enhancement of miotic potential of pilocarpine by tamarind gum based *in situ* gelling ocular dosage form. Acta Pharm. Sci. **52**, 145–154 (2010)

M.U. Mishra, J.N. Khandare, Evaluation of tamarind seed polysaccharide as a biodegradable carrier for colon specific drug delivery. Int. J. Pharm. Pharmaceut. Sci. **3**, 139–142 (2011)

R.C. Nandi, A process for preparation of polyose from the seeds of *Tamarindus indica*, Indian Patent 142092 (1975)

A.K. Nayak, Advances in therapeutic protein production and delivery. Int. J. Pharm. Pharmaceut. Sci. **2**, 1–5 (2010)

A.K. Nayak, M.S. Hasnain, D. Pal (2018) Gelled microparticles/beads of sterculia gum and tamarind gum for sustained drug release, in *Handbook of Springer on Polymeric Gel*, ed. V.K. Thakur, M.K. Thakur (Springer International Publishing, Switzerland), pp. 361–414

A.K. Nayak, D. Pal, Development of pH-sensitive tamarind seed polysaccharide-alginate composite beads for controlled diclofenac sodium delivery using response surface methodology. Int. J. Biol. Macromol. **49**, 784–793 (2011)

A.K. Nayak, D. Pal, Ionotropically-gelled mucoadhesive beads for oral metformin HCl delivery: formulation, optimization and antidiabetic evaluation. J. Sci. Ind. Res. **72**, 15–22 (2013)

A.K. Nayak, D. Pal, Tamarind seed polysaccharide: an emerging excipient for pharmaceutical use. Indian J. Pharm. Educ. Res. **51**, S136–S146 (2017)

A.K. Nayak, D. Pal, J. Malakar, Development, optimization and evaluation of emulsion-gelled floating beads using natural polysaccharide-blend for controlled drug release. Polym. Eng. Sci. **53**, 338–350 (2013)

A.K. Nayak, D. Pal, K. Santra, Development of calcium pectinate-tamarind seed polysaccharide mucoadhesive beads containing metformin HCl. Carbohydr. Polym. 101, 220–230 (2014)

A.K. Nayak, D. Pal, K. Santra, Tamarind seed polysaccharide-gellan mucoadhesive beads for controlled release of metformin HCl. Carbohydr. Polym. **103**, 154–163 (2014)

A.K. Nayak, D. Pal, K. Santra, Swelling and drug release behavior of metformin HCl-loaded tamarind seed polysaccharide-alginate beads. Int. J. Biol. Macromol. **82**, 1023–1027 (2016)

D. Pal, A.K. Nayak, Novel tamarind seed polysaccharide-alginate mucoadhesive microspheres for oral gliclazide delivery. Drug Deliv. **19**, 123–131 (2012)

B. Patel, P. Patel, A. Bhosale, S. Hardikar, S. Mutha, G. Chaulang, Evaluation of tamarind seed polysaccharide (TSP) as a mucoadhesive and sustained release component of nifedipine buccoadhesive tablet & comparison with HPMC and Na CMC. Int. J. PharmTech Res. **1**, 404–410 (2009)

G.K. Phani Kumar, G. Battu, I. Raju, N.S. Kotha, Studies on the applicability of tamarind seed polymer for the design of sustained release matrix tablet of tramadol HCl. J. Pharm. Res. **4**, 703–705 (2011)

P.S. Rao, T.P. Ghosh, S. Krishna, Extraction and purification of tamarind seed polysaccharide. J. Sci. Ind. Res. **4**, 705 (1946)

P.S. Rao, H.C. Srivastava, Tamarind, in *Industrial gums*, 2nd edn., ed. by R.L. Whistler (Academic Press, New York, 1973), pp. 369–411

M. Rishabha, S. Pranati, K. Upendra, C.S. Bhargava, S.P. Kumar, Formulation and comparison of suspending properties of different natural polymers using paracetamol suspension. Int. J. Drug. Dev. Res. **2**, 886–891 (2010)

S. Sahoo, R. Sahoo, P.L. Nayak, Tamaind seed polysaccharide: a versatile biopolymer for mucoadhesive applications. J. Pharm. Biomed. Sci. **8**, 1–12 (2010)

P.K. Samal, J.S. Dangi, Isolation, preliminary characterization and hepatoprotective activity of polysaccharides from *Tamarindus indica* L. Carbohydr. Polym. **102**, 1–7 (2014)

S. Sumathi, A.R. Ray, Release behavior of drugs from tamarind seed polysaccharide tablets. J. Pharm. Pharmaceut. Sci. **5**, 12–18 (2002)

J. Tattiyakul, C. Muangnapoh, S. Poommarinvarakul, Isolation and rheological properties of tamarind seed polysaccharide from tamarind karnel powder using protease enzyme and high-intensity ultrasound. J. Food Sci. **75**, 253–260 (2010)

J. Zhang, S. Xu, S. Zhang, Z. Du, Preparation and characterization of tamarind gum/sodium alginate composite gel beads. Iran. Polym. J. **17**, 899–906 (2008)

Chapter 5
Locust Bean Gum Based Multiple Units for Oral Drug Delivery

5.1 Locust Bean Gum (LG)

Locust/carob bean gum (LG) is a branched biopolysaccharide derived from plant which is nonionic in nature and obtained from the seeds of carob tree (*Ceratonia siliqua*) (Malik et al. 2011; Kaity et al. 2013). It is a galactomannan type polysaccharide which contains galactose and mannose in a ratio of 1: 4 (Parvathy et al. 2005). It consists of (1, 4)-linked β-d-mannopyranose skeleton with branch points at six positions and is connected to α-d-galactose (Kaity et al. 2013). LG is less water-soluble and needed heating to get a ready aqueous solution (Dionísio and Grenha 2012). LG possesses the ability to produce very viscous solutions at the lower concentrations and that is more or less unaltered by the addition of salts, pH changes, or temperature changes (Mathur and Mathur 2005). LG is biocompatible as well as biodegradable polysaccharide material. It is generally known as nonteratogenic and nonmutagenic (Prajapati et al. 2013a).

5.2 Use of LG as Pharmaceutical Excipients

LG is employed as natural polymeric pharmaceutical excipients in several kinds of pharmaceutical dosage forms (Dionísio and Grenha 2012; Prajapati et al. 2013a, b). It is used to formulate tablets of oral administration. As LG has been reported to have mucoadhesive profile, it is utilized as mucoadhesive polymer in the formula of buccal patches. LG is also investigated as the polymeric blends with sodium alginate to formulate the ionotropically gelled microparticles of LG-blended alginate and beads for sustained drug release (Prajapati et al. 2014, 2015a). Pharmaceutical applications of LG in different formulations are summarized in Table 5.1.

© The Author(s), under exclusive license to Springer Nature Singapore Pte Ltd. 2019   61
A. K. Nayak and M. S. Hasnain, *Plant Polysaccharides-Based Multiple-Unit
Systems for Oral Drug Delivery*, SpringerBriefs in Applied Sciences and Technology,
https://doi.org/10.1007/978-981-10-6784-6_5

Table 5.1 Pharmaceutical applications of LG in different formulations

Formulations made of LG	Pharmaceutical applications	References
Nimesulide tablets	Superdisintegrant	Malik et al. (2011)
Matrix tablets of propranolol HCl	Matrix former, release retardant	Venkataraju et al. (2007)
Matrix tablet of theophylline and myoglobin	Matrix former, release retardant	Coviello et al. (2007)
LG/xanthan gum hydrogel based compressed tablets of myoglobin	Matrix former, release retardant	Sandolo et al. (2007)
Alginate–LG IPN microspheres of aceclofenac	Matrix former, release retardant	Jana et al. (2015)
Alginate–LG mucoadhesive beads of aceclofenac	Matrix former, release retardant	Prajapati et al. (2014)
Alginate–LG beads of gliclazide	Matrix former, release retardant	Prajapati et al. (2013c)
Controlled release alginate microspheres using LG	Matrix former, release retardant	Deshmukh et al. (2009)
Tablets made of IPN of LG-poly (vinyl alcohol)	Matrix former, release retardant	Kaity et al. (2013)
Sodium diclofenac mini-matrices	Matrix former, release retardant	Sujja-areevath et al. (1998)
Metoprolol tartrate buccoadhesive tablets containing LG or a mixture of LG and xanthan gum	Matrix former, release retardant, mucoadhesive	Yamagar et al. (2010)
Prednisolone releasing hydrogels	Matrix former, release retardant	Watanabe et al. (1992)

5.3 LG–Alginate Microspheres of Aceclofenac

Jana et al. (2015) prepared LG–alginate microspheres for sustained releasing of aceclofenac via calcium-ion-induced ionotropically gelation method. The percentage
yield of the prepared LG–alginate microspheres of aceclofenac was $64.44 \pm 2.18\%$
to $91.64 \pm 2.49\%$, whereas the drug entrapment efficiency of these aceclofenac-
encapsulated microspheres was reported $59.64 \pm 2.60\%$ to $93.25 \pm 1.09\%$, respectively. The utmost yield could be obtained due to LG and alginate insolubility in calcium chloride solutions (ionotropic cross-linking solution) and this might possibly
lose the negligible quantity of drug (here aceclofenac) in of calcium chloride solutions during ionotropic gelation. Better drug entrapment efficiency of these LG–alginate microspheres of aceclofenac might be linked with the concentrations of sodium
alginate used in the polymeric blends and the potential of cross-linking calcium
ions with anionic sodium alginate for ionotropic cross-linking. The aceclofenac-
encapsulated LG–alginate microspheres of higher efficiency of aceclofenac

entrapment were found to be produced as the concentrations of calcium chloride solutions in the ionotropic cross-linking solution which was augmented during preparation. The formulated aceclofenac-encapsulated LG–alginate microspheres were relatively spherical in shape and sizes of the LG–alginate microspheres were within a range of 406 ± 10.18 µm to 684 ± 23.36 µm. In vitro swelling index of the formulated aceclofenac-encapsulated LG–alginate microspheres acidic milieu (pH 1.2) was reported in a range of $26 \pm 1.2\%$ to $44 \pm 3.2\%$, whereas in the alkaline milieu (pH 7.4), it was $168 \pm 4.6\%$ to $288 \pm 1.6\%$. The swelling results indicated that the swelling of the microspheres was lesser in a gastric pH milieu in comparison to that of intestinal pH while swelling of ionotropically gelled alginate-based systems occurred in the smaller quantity in the acidic milieu. The lesser swelling behavior of these prepared alginate-based ionotropically gelled hydrogel systems in the acidic milieu was most likely because of proton–calcium ion exchange producing insoluble regions of alginic acid followed by the incursion of the solvent into the network of ionotropically gelled alginate hydrogel. The FTIR spectroscopy results indicated that no interaction was found in between drug and polymers within the prepared ionotropically gelled aceclofenac-encapsulated LG–alginate microspheres. Study of in vitro drug releasing of the prepared aceclofenac-encapsulated LG–alginate microspheres were performed in phosphate buffer at a pH of 6.8 and observed to be sustained over 8 h. As the polymer concentration (for both sodium alginate and LG) augmented in the polymeric blends, which were employed to prepare aceclofenac-encapsulated LG–alginate microspheres, the aceclofenac release rate of these microspheres was found to be decreased. The data of in vitro aceclofenac releasing of LG–alginate microspheres were shown to be best fitted in Korsmeyer–Peppas model over 8 h. Non-Fickian (anomalous diffusion mechanism) mechanism of drug releasing was also envisaged. The formulated optimized ionotropically gelled aceclofenac-encapsulated LG–alginate microspheres were evaluated in the rat paw oedema model (carrageenan induced) for their pharmacodynamic activity. A prolonged anti-inflammatory activity over 5 h in carrageenan-treated rats was reported after the oral administration of aceclofenac-encapsulated LG–alginate microspheres.

5.4 LG–Alginate Mucoadhesive Beads of Aceclofenac

Prajapati et al. (2014), in another study, developed aceclofenac-encapsulated mucoadhesive beads of LG–alginate by ionotropic-gelation method using calcium chloride as the cross-linking agent with the idea of expanding the utilization of LG in the development of mucoadhesive multiple unit for the sustained drug release systems. The 3^2-factorial design was employed with the aid of a computer for the optimization of LG–alginate mucoadhesive beads. The effects of LG and sodium alginate quantities as polymeric blends on drug entrapment, mucoadhesivity after 8 h and in vitro cumulative drug releasing after 10 h were explored and optimized. The yields of the prepared LG–alginate beads of aceclofenac were observed in a range of 93.19–96.65%, while drug entrapment efficiency of these prepared beads was reported in the range of 56.37–68.54%. The results of drug entrapment study showed that LG did not demonstrate any influence on drug entrapment of these prepared aceclofenac-encapsulated LG–alginate beads. As the amount of sodium alginate in beads increased, the drug entrapment within these aceclofenac-encapsulated LG–alginate beads was observed to be enhanced. Average sizes of these prepared aceclofenac-encapsulated LG–alginate beads were measured in the range 1.32 ± 0.11 μm to 1.42 ± 0.13 μm. The developed beads were of uniform in size, distinct, and free flowing. Images of scanning electron microscopy (SEM) demonstrated that the prepared beads were spherical in shape. Fourier-transform infrared (FTIR) and differential scanning calorimetry (DSC) analyses recommended the compatibility in between encapsulated drug (aceclofenac) and excipient. The pH of swelling media influenced the swelling of these prepared aceclofenac-encapsulated LG–alginate beads and this did not exhibit dissolution of polymer matrix at pH 1.2 and 6.8 (i.e., gastric pH milieu and intestinal pH milieu, respectively) over 10 h of in vitro swelling. These prepared aceclofenac-loaded beads showed markedly lower swelling ratio at pH 1.2 as compared to that at pH 6.8. Again, it was observed that the swelling ratio of the prepared beads was augmented due to rise in concentration of polymers in aceclofenac-encapsulated LG–alginate beads. The wash-off study was performed for the measurement of mucoadhesiveness of these prepared aceclofenac-encapsulated LG–alginate beads at both pH 1.2 and 6.8 for 10 h by means of intestinal mucosal tissue of goat. In gastric pH milieu, these beads hold onto the mucosal tissue of intestine and were reported to vary in the range of 30.00 ± 1.29% to 45.00 ± 1.76% over 8 h. Furthermore, this was varied in a range of 60 ± 1.62% to 85 ± 1.83% in intestinal pH milieu. Consequently, the prepared aceclofenac-encapsulated LG–alginate beads exhibited excellent mucoadhesiveness in the intestinal pH milieu onto the mucosal tissue. The in vitro drug release from the beads was performed in gastric pH milieu for the initial 2 h and for next 10 h in intestinal pH milieu. Sustained drug release was observed over 12 h of drug release study and also found depended on the compositions of polymeric blends (coat polymers). Rate of in vitro drug release of the prepared aceclofenac-encapsulated LG–alginate beads was reported reasonably slower in gastric pH milieu with respect to that of in intestinal pH milieu. Release of aceclofenac from the prepared aceclofenac-encapsulated beads (which were reported as 84.95 ± 2.02% to 95.33 ± 1.56% at 10 h) showed a sustained drug release pattern (first order) with super case-II transport drug-releasing mechanism over 12 h. Further, it was observed that the release of drug from the formulated ionotropically gelled polymeric systems increases as the sodium alginate quantity increases in the LG–sodium alginate polymeric blends which were employed in the formulation of these beads, which was observed as a result of release rate retardant action of sodium alginate. Release of drug from these formulated aceclofenac-encapsulated LG–alginate beads was not apparent in acidic milieu, whereas it was highest in alkaline milieu.

5.5   Carboxymethyl LG Beads of Glipizide

Maiti et al. (2010) synthesized carboxymethyl LG and developed glipizide-encapsulated ionotropically gelled beads using chemical synthesized carboxymethyl LB by means of varying the concentration of cross-linker (aluminum chloride). These formulated carboxymethyl LG beads of glipizide were of spherical shaped and the drug (glipizide) entrapment efficiency was measured within the range of 85.14–97.68%. Drug entrapment was found to be decreased as the concentration of aluminum chloride increased. Release of encapsulated drug (glipizide) from these formulated carboxymethyl LG beads of glipizide was found relatively slower in the acidic milieu with respect to the alkaline milieu. The release of encapsulated drug was observed slower for these beads which were prepared with higher concentration of cross-linking solution. In addition, these carboxymethyl LG beads of glipizide indicated an encouraging prolonged hypoglycemic effect.

References

T. Coviello, F. Alhaique, A. Dorigo, P. Matricardi, M. Grassi, Two galactomannans and scleroglucan as matrices for drug delivery: preparation and release studies. Eur. J. Pharm. Biopharm. **66**, 200–209 (2007)

V.N. Deshmukh, J.K. Jadhav, V.J. Masirkar, D.M. Sakarkar, Formulation, optimization and evaluation of controlled release alginate microspheres using synergy gum blends. Res. J. Pharm. Tech. **2**, 324–327 (2009)

M. Dionísio, A. Grenha, Locust bean gum: Exploring its potential for biopharmaceutical applications. J. Pharm. Bioall. Sci. **4**, 175–185 (2012)

S. Jana, A. Gangopadhaya, B.B. Bhowmik, A.K. Nayak, A. Mukhrjee, Pharmacokinetic evaluation of testosterone-loaded nanocapsules in rats. Int. J. Biol. Macromol. **72**, 28–30 (2015a)

S. Jana, A. Gandhi, S. Sheet, K.K. Sen, Metal ion-induced alginate–locust bean gum IPN microspheres for sustained oral delivery. Int. J. Biol. Macromol. **72**, 47–53 (2015b)

S. Kaity, J. Isaac, A. Ghosh, Interpenetrating polymer network of locust bean gum-poly (vinyl alcohol) for controlled release drug delivery. Carbohydr. Polym. **94**, 456–467 (2013)

K. Malik, G. Arora, I. Singh, Locust bean gum as superdisintegrant – formulation and evaluation of nimesulide. Polym. Med. **41**, 17–28 (2011)

V. Mathur, N. Mathur, Fenugreek and other less known legume galactomannan polysaccharides: scope for developments. J. Sci. Ind. Res. **64**, 475–481 (2005)

S. Maiti, P. Dey, A. Banik, B. Sa, S. Ray, S. Kaity, Tailoring of locust bean gum and development of hydrogel beads for controlled oral delivery of glipizide. Drug Deliv **17**, 288–300 (2010)

K.S. Parvathy, N.S. Susheelamma, R.N. Tharanathan, A.K. Gaonkar, A simple non-aqueous method for carboxymethylation of galactomannans. Carbohydr. Polym. **62**, 137–141 (2005)

V.D. Prajapati, G.K. Jani, N.G. Moradiya, N.P. Randeria, Pharmaceutical applications of various natural gums, mucilages and their modified forms. Carbohydr. Polym. **92**, 1685–1699 (2013a)

V.D. Prajapati, G.K. Jani, N.G. Moradiya, N.P. Randeria, B.J. Nagar, Locust bean gum: a versatile biopolymer. Carbohydr. Polym. **94**, 814–821 (2013b)

V.D. Prajapati, K.H. Mashuru, H.K. Solanki, G.K. Jani, Development of modified release gliclazide biological macromolecules using natural biodegradable polymers. Int. J. Biol. Macromol. **55**, 6–14 (2013c)

V.D. Prajapati, G.K. Jani, N.G. Moradiya, N.P. Randeria, P.M. Maheriya, B.J. Nagar, Locust bean gum in the development of sustained release mucoadhesive macromolecules of aceclofenac. Carbohydr. Polym. **113**, 138–148 (2014)

C. Sandolo, T. Coviello, P. Matricardi, F. Alhaique, Characterization of polysaccharide hydrogels for modified drug delivery. Eur. Biophys. J. **36**, 693–700 (2007)

J. Sujja-areevath, D. Munday, P. Cox, K. Khan, Relationship between swelling, erosion and drug release in hydrophilic natural gum mini-matrix formulations. Eur. J. Pharm. Sci. **6**, 207–217 (1998)

M.P. Venkataraju, D.V. Gowda, K.S. Rajesh, K.H. Shiva, Xanthan and locust bean gum matrix tablets for oral controlled delivery of propranolol HCl. Asian J. Pharm. Sci. **2**, 239–248 (2007)

K. Watanabe, S. Yakou, K. Takayama, Y. Machida, T. Nagai, Factors affecting prednisolone release from hydrogels prepared with water-soluble dietary fibers, xanthan and locust bean gums. Chem. Pharm. Bull. **40**, 459–462 (1992)

M. Yamagar, V. Kadam, R. Hirlekar, Design and evaluation of buccoadhesive drug delivery system of metoprolol tartrate. Int. J. PharmTech Res. **2**, 453–462 (2010)

Chapter 6
Sterculia Gum Based Multiple Units for Oral Drug Delivery

6.1 Sterculia Gum (SG)

SG is generally called as karaya gum (Gauthami and Bhat 1992; Nayak et al. 2018). It is a plant-derived polysaccharide having medicinal importance. It has a high molecular weight and is water soluble (Nayak and Pal 2016). It is found from the exudates of *Sterculia urens* plant, belonging to the family: Sterculiaceae (Cerf et al. 1990; Nayak et al. 2018). The crude form of gum is usually obtained as exudates by peeling or cutting the *Sterculia urens* tree bark. SG is a partly acetylated polysaccharide and made up of three dissimilar chains of polysaccharide (Singh and Chauhan 2011; Singh et al. 2010). The first chain (i.e., fifty percent of the total polysaccharide) consists of four galacturonic acid residues repeating units, having residues of L-rhamnose at the reducing end and β-D-galactose branch. The second chain (i.e., seventeen percent of the total polysaccharide) consists of an oligorhamnan containing D-galactose residues and D-galacturonic acid branch residues. The last or 3rd chain (i.e., thirty percent of the total polysaccharide) has D-glucuronic acid residues comprising about galactose (13–26%), rhamnose (15–30%), and uronic acid residue (40%) (Cerf et al. 1990; Kulkarni et al. 2014). SG possesses a number of distinctive characteristics such as good viscosity, greater acidic stability, and high swelling capability (Cerf et al. 1990; Silva et al. 2003; Singh and Sharma 2011). In the USA, SG is ranked as "Generally Recognized as Safe" ("GRAS") (Anderson 1989; Singh et al. 2010). It is reported to be nonallergic, nonteratogenic, nonmutagenic, and nontoxic nature (Gauthami and Bhat 1992; Singh and Sharma 2011; Singh et al. 2011). SG also possesses antimicrobial characteristics (Gauthami and Bhat 1992; Singh et al. 2010, 2011). It improves glucose metabolism and decreases cholesterol level without any affecting the mineral balance (Behall et al. 1987). SG has been used as a medication for diarrhea (Huttel 1983), ulcers (Zide and Bevin 1980), irritable bowel syndrome (Capron et al. 1981) and chronic colonic disorders (Guerre and Neuman 1979) and as

© The Author(s), under exclusive license to Springer Nature Singapore Pte Ltd. 2019
A. K. Nayak and M. S. Hasnain, *Plant Polysaccharides-Based Multiple-Unit Systems for Oral Drug Delivery*, SpringerBriefs in Applied Sciences and Technology,
https://doi.org/10.1007/978-981-10-6784-6_6

laxative (Meier et al. 1990). SG is used as emulsifier, stabilizer, and as food thickener excipient (Anderson and Wang 1994). Because of its analogous physical properties, it is often adulterated with Gum tragacanth (Weiping 2000).

6.2 Use of SG as Pharmaceutical Excipients

From the last century, SG is acknowledged as a potential polysaccharide biomaterial, which is biodegradable. It is exploited in the formulation of a variety of drug delivery dosages (Deshmukh et al. 2009; Nayak and Pal 2016; Nayak et al. 2018; Singh and Chauhan 2011). SG is exploited as a useful pharmaceutical excipient material in tablets (Deshmukh et al. 2009; Park and Munday 2004), microparticles (Kulkarni et al. 2014), beads (Bera et al. 2015b; Guru et al. 2013), hydrogels (Singh and Sharma 2011), buccoadhesive drug delivery systems (Bera et al. 2015b), etc. Pharmaceutical applications of SG in different formulations are summarized in Table 6.1.

6.3 Oil-Entrapped SG–Alginate Floating Beads of Aceclofenac

Guru et al. (2013) investigated the formulation development of oil-entrapped float-ing beads for gastroretentive delivery of aceclofenac exploiting SG–sodium alginate blends. These oil-entrapped SG–alginate floating beads of aceclofenac were prepared via the methodology of ionotropic emulsion gelation using aqueous calcium chlo-ride solution as ionotropic cross-linking solution. In order to examine the impacts of independent process variables such as polymer to drug ratio and sodium algi-nate to SG ratio on the drug entrapment and the drug releasing of oil-entrapped SG–alginate floating beads of aceclofenac, a 3^2-factorial design-assisted optimiza-tion technique was employed. Three-dimensional response surface graphs and two-dimensional response surface graphs showing the impacts of sodium alginate to SG ratio and polymer to drug ratio on (a) drug entrapment efficiency (%) and (b) cumu-lative drug release after 7 h (%) in simulated gastric fluid (pH 1.2) are presented in Figs. 6.1 and 6.2, respectively. The aceclofenac entrapment efficiency of these oil-entrapped SG–alginate floating beads of aceclofenac ranged from 63.28 ± 0.55% to 90.92 ± 2.34%, while the optimized formulation exhibited 83.73 ± 0.81% of aceclofenac entrapment. As sodium alginate to SG ratio was decreased and the poly-mer to aceclofenac ratio was increased, the efficiency of aceclofenac entrapment of these SG–alginate beads was found augmented. The increment of polymer to ace-clofenac ratio in these oil-entrapped SG–alginate beads of aceclofenac could form entanglements of higher aceclofenac quantity inside the cross-linked SG–alginate gel network. This could help to facilitate the drug entrapment increment. The mean sizes of these SG–alginate beads were within 1.32 ± 0.04 to 1.72 ± 0.12 mm and it

Table 6.1 Pharmaceutical applications of SG in different formulations

Formulations made of SG	Pharmaceutical applications	References
Matrix tablet for sustained drug release	Matrix former, release retardant	Sreenivasa et al. (2000)
Compressed tablets made of xanthan gum and SG	Matrix former, release retardant	Munday and Philip (2000)
Diclofenac-sodium-controlled-release tablets prepared from SG–chitosan polyelectrolyte complexes	Matrix former, release retardant	Lankalapalli and Kolapalli (2012)
Theophylline anhydrous bioadhesive tablets	Matrix former, release retardant, mucoadhesive	Deshmukh et al. (2009)
Buccoadhesive tablets for sustained release of nicotine	Matrix former, release retardant, mucoadhesive	Park and Munday (2004)
SG–alginate IPN microparticles of repaglinide	Encapsulating material and release retardant	Kulkarni et al. (2014)
Mucoadhesive-floating zinc–pectinate–SG IPN beads of ziprasidone HCl	Encapsulating material and release retardant	Bera et al. (2015b)
Alginate–SG gel-coated oil-entrapped alginate beads of risperidone	Encapsulating material and release retardant	Bera et al. (2015)
Oil-entrapped SG–alginate buoyant systems of aceclofenac	Encapsulating material and release retardant	Guru et al. (2013)
SG–calcium alginate floating and non-floating beads of pentoprazole	Encapsulating material and release retardant	Singh et al. (2010)
SG based hydrogel system double potential antidiarrhoeal drug delivery system	Matrix former, release retardant	Singh and Sharma (2011)
Novel hydrogels by modification of SG through radiation cross-linking polymerization for use in drug delivery	Matrix former, release retardant	Singh and Vashishtha (2008)
Novel hydrogels by modification of SG with methacrylic acid for use in drug delivery	Matrix former, release retardant	Singh and Sharma (2008)
SG cross-linked PVA and PVA-poly(AAm) hydrogel wound dressings for slow drug delivery	Matrix former, release retardant, mucoadhesive, film-former	Singh and Pal (2012)

was 1.62 ± 0.08 mm for the optimized beads. The densities of these oil-entrapped SG–alginate beads of aceclofenac were calculated less than the gastric fluid density, imparting floatation (buoyancy). All these oil-entrapped SG–alginate beads of aceclofenac were also seen to buoyant within 6 min after being placed in the acidic milieu (simulated gastric fluid, pH 1.2) and demonstrated buoyancy over a period of 7 h. The entrapment of low-density oil like liquid paraffin within SG–alginate floating beads of aceclofenac was accountable to impart low density and floatation of oil-entrapped polymeric beads contained aceclofenac. Scanning electron microscopy (SEM) results of the optimized oil-entrapped SG–alginate floating beads of aceclofenac demonstrated a rough surface morphology having corrugations, channels or/and small pores (Fig. 6.3). Fourier-transform infrared (FTIR) spectroscopy studies revealed that the compatibility of the encapsulated drug with polymers used (SG and sodium alginate) to prepare oil-entrapped SG–alginate floating beads of aceclofenac. X-ray diffraction (XRD) results demonstrated the crystalline character of the pure drug (aceclofenac). On the other hand, the crystalline character of the encapsulated drug was significantly decreased in the optimized oil-entrapped SG–alginate floating beads of aceclofenac. This may possibly be attributable to the influence of the polymers used (SG–alginate) or the formulation procedure exploited. In vitro aceclofenac releasing from these oil-entrapped SG–alginate floating beads of aceclofenac in the medium of gastric pH (1.2), which clearly demonstrated a sustained drug releasing pattern over 7 h (Fig. 6.4). The increment in the in vitro aceclofenac releasing from these oil-entrapped SG–alginate floating beads was detected as the polymer to drug ratio increased. The slow in vitro aceclofenac releasing was also detected as the oil (liquid paraffin) entrapment increased in these biopolymeric floating beads, which can be explicated as the majority of the encapsulated drug amounts remained saturated as well as dispersed in the oil pockets of the oil-entrapped SG–alginate floating beads to structure an oil-drug dispersed matrix system. This also indicated controlled releasing of encapsulated aceclofenac with the Korsmeyer–Peppas model kinetics and the anomalous (non-Fickian) release mechanism over 7 h.

6.4 SG–Alginate Floating and Non-floating Beads of Pentoprazole

As SG and alginates possess a therapeutic potential to treat ulcers, Singh et al. (2010) formulated SG–alginate non-floating beads and floating beads of pentoprazole. Both the SG–alginate based non-floating beads and floating beads of pentoprazole were prepared via the ionotropic-gelation by means of ionotropic cross-linking solutions of calcium chloride. To formulate SG–alginate floating beads of pentoprazole, the amalgamation of calcium carbonate (2% w/v as effervescent agent) was employed. The calcium cations caused the ionotropic gelation in a medium of pH less than 7 and this assisted in forming a gelled barrier on the prepared floating bead surface. Actually, the effervescent material calcium carbonate liberated calcium cations and

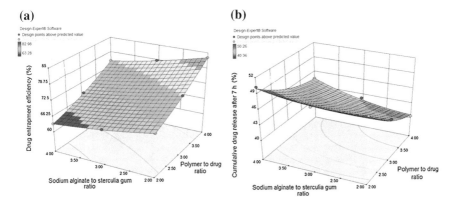

Fig. 6.1 Three-dimensional response surface graphs showing the impacts of sodium alginate to SG ratio and polymer to drug ratio on **a** drug entrapment efficiency (%) and **b** cumulative drug release after 7 h (%) in simulated gastric fluid (pH 1.2) [Guru et al. (2013); Copyright @ 2013, with permission from Elsevier B.V.]

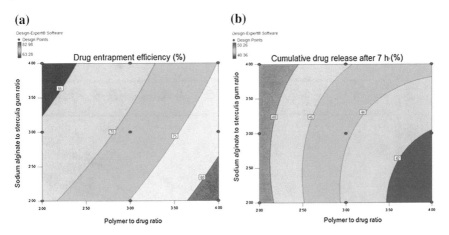

Fig. 6.2 Two-dimensional response surface graphs showing the impacts of sodium alginate to SG ratio and polymer to drug ratio on **a** drug entrapment efficiency (%) and **b** cumulative drug release after 7 h (%) in simulated gastric fluid (pH 1.2) [Guru et al. (2013); Copyright @ 2013, with permission from Elsevier B.V.]

carbon dioxide. The divalent calcium cations interacted with the carboxyl groups of sodium alginate and SG (as both polymers are anionic in nature) forming a three-dimensional gelled (ionotropically) network structure confining the diffusion of carbon dioxide. Both the SG–alginate based non-floating beads of pentoprazole were of spherically shaped. The surface morphological investigation of SG–alginate non-floating beads and floating beads of pentoprazole was carried out by SEM analyses. The SEM image of SG–alginate non-floating beads of pentoprazole demonstrated a rough surface morphological feature. The SEM image of SG–alginate floating

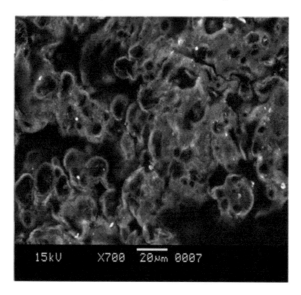

Fig. 6.3 SEM photograph of optimized oil-entrapped SG–alginate floating beads of aceclofenac showing a rough surface with pores and channels [Guru et al. (2013); Copyright @ 2013, with permission from Elsevier B.V.]

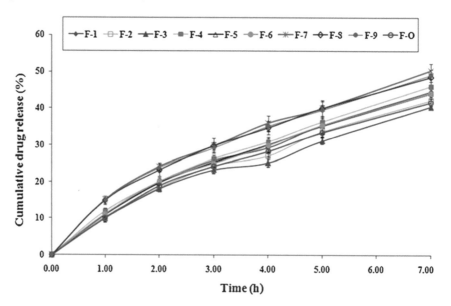

Fig. 6.4 In vitro aceclofenac releasing from these oil-entrapped SG–alginate floating beads of aceclofenac in the medium of gastric pH (1.2) [Guru et al. (2013); Copyright @ 2013, with permission from Elsevier B.V.]

beads of pentoprazole demonstrated a comparatively smooth surface morphology. The releasing of divalent calcium cations from calcium carbonate may possibly form consistent SG–alginate floating beads of pentoprazole. This occurrence could attribute for the smoother surface morphology of the SG–alginate floating beads of pentoprazole. Moreover, this could be attributed to the higher bursting impacts of carbon dioxide liberation to harden the walls of these SG–alginate based beads. The electron dispersion X-ray (EDX) analyses demonstrated the occurrence of carbon, hydrogen and oxygen within these SG–alginate beads. These three major elements comprise the chemical structure constituents of the polysaccharide. EDX results also demonstrated that the occurrence of calcium cation cross-linked ionotropic-gelation in the SG–alginate non-floating and floating beads of pantoprazole. The analysis performed by FTIR demonstrated that mutually these SG–alginate non-floating and floating beads of pentoprazole had typical characteristic peaks of SG as well as that of alginate. The in vitro swelling behavior of SG–alginate non-floating and float-ing beads of pentoprazole indicated the impact of calcium chloride concentration (ionotropic cross-linker), the quantity of sodium alginate, the quantity of SG, and pH of the swelling medium. These SG–alginate floating beads demonstrated the floating performance for a longer duration. The loading of drug (pentoprazole) in these biopolymeric floating and non-floating beads of pentoprazole was 66.10% and 83.90%, respectively. Sustained in vitro releasing of encapsulated pentoprazole from SG–alginate non-floating and floating beads of pentoprazole over 24 h at buffer pH, 2.2 and in distilled water was observed. These SG–alginate beads of pentoprazole indicated that the drug releasing followed by Fickian diffusion mechanism.

In another work, Singh and Chauhan (2011) formulated the same type of SG–al-ginate floating beads of pentoprazole using barium chloride as ionotropic cross-linker. The loading of pentoprazole in these barium-ion-induced ionotropically gelled SG–alginate non-floating and floating of pentoprazole were 61.60% and 67.90%, respectively. The in vitro swelling pattern of these SG–alginate floating beads of pentoprazole was found increased in pH 7.4 buffer as compared to that of in distilled water and in pH 2.2 buffer. A reduced in vitro swelling pattern was calculated in case of the barium-ion-induced SG–alginate floating beads of pentoprazole as compared to that of SG–alginate non-floating beads of pentoprazole. Simultaneously, these barium-ion-induced SG–alginate floating beads of pentoprazole demonstrated more stability in pH 7.4 buffers as compared to that of calcium-ion-induced SG–alginate floating beads of pentoprazole. Since barium cations and calcium cations both are divalent cations, the ionotropic cross-linkings with alginate by barium cations and calcium cations are supposed to take place in a planner two-dimensional manner inside these SG–alginate beads. Though, barium cations encompass the larger radius of 1.74 Å than the radius of calcium cations (1.14 Å). Therefore, it is expected to fill up a larger gap within the cross-linking polymer chains that generates a rigid arrangement with the comparatively smaller voids. In these beads, the exchange of the comparatively larger divalent barium cations in these SG–alginate beads with the monovalent sodium ions and also their removal in the form of insoluble barium phos-phate could be hindered. This could produce an impact on the lowest water uptake as well as the higher degree of stability. The in vitro pentoprazole releasing from these

SG–alginate floating beads of pentoprazole was found to be sustained over 24 h in a different medium for drug release such as distilled water, pH 2.2 buffers, and pH 7.4 buffers. The in vitro pentoprazole releasing from these SG–alginate floating beads followed the Fickian diffusion mechanism.

6.5 SG–Alginate Gel-Coated Oil-Entrapped Alginate Beads of Risperidone

Bera et al. (2015b) developed SG–alginate gel-coated oil-entrapped alginate beads of risperidone for the use in gastroretentive drug delivery using the combination mechanism of mucoadhesion–floatation. For the formulation development of SG–alginate gel-coated oil-entrapped alginate beads of risperidone, 1% w/w SG–sodium alginate aqueous dispersion in a ratio of 1:1 and next, transferred into 5% w/v calcium chloride (ionotropic cross-linker) solution for a period of 10 min to harden. The SG–alginate gel-coated beads were washed thoroughly using distilled water and subsequently, dried at room temperature for overnight. These SG–alginate gel-coated oil-entrapped alginate beads possessed $81.63 \pm 1.54\%$ of risperidone entrapment efficiency, 2.49 ± 0.12 mm of mean diameter and 0.66 ± 0.15 g/cm^3 of density. SEM analyses of SG–alginate gel-coated oil-entrapped alginate beads of risperidone demonstrated spherical shaped beads with a comparative smoother surface in comparison with that of the optimized uncoated beads (Fig. 6.5a, b). The coated membrane of SG–alginate gel onto the oil-entrapped beads could lessen the cracks as well as pores on the bead surface. SEM photograph of the cross-sectional view of the SG–alginate gel-coated oil-entrapped alginate beads of risperidone demonstrated a sponge-like structural morphology, indicating oil-entrapment occurrence (Fig. 6.5c, d). The SG–alginate gel-coated oil-entrapped alginate beads of risperidone demonstrated sustained in vitro releasing of risperidone in the gastric pH (1.2) over 8 h (Fig. 6.6). The in vitro releasing results demonstrated a slower release pattern than the uncoated oil-entrapped alginate beads of risperidone. The SG–alginate gel-coated membrane onto the oil-entrapped alginate beads of risperidone could be worked as a hydrophilic gelled barrier responsible for retarding risperidone releasing, in vitro. In addition, the SG–alginate gel coating could have been produced blocking the pores onto the surface of oil-entrapped alginate beads of risperidone. This occurrence also could lead to slow the rate of drug releasing. The developed SG–alginate gel-coated oil-entrapped alginate beads of risperidone obeyed Korsmeyer–Peppas model with Fickian diffusion mechanism of drug releasing. These SG–alginate gel-coated oil-entrapped alginate beads of risperidone exhibited excellent buoyancy with negligible floating lag-time (Fig. 6.7a), higher swelling (Fig. 6.7b), and high-quality biomucoadhesion onto the gastric mucosal surface (Fig. 6.7c) in comparison with that of the uncoated beads contained risperidone.

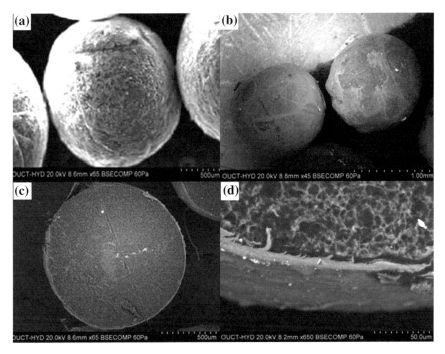

Fig. 6.5 SEM photograph of the uncoated and SG–alginate gel-coated oil-entrapped alginate beads of risperidone showing rough surface of the uncoated beads (**a**), smooth surface of the coated beads (**b**), cross-sectional view of the coated beads with less zoom (65 ×) (**c**) and high zoom (650 ×) (**d**) [Bera et al. (2015c); Copyright @ 2014, with permission from Elsevier Ltd.]

6.6   SG–Pectinate Floating-Mucoadhesive IPN Beads of Ziprasidone HCl

Bera et al. (2015b) developed SG–pectinate floating-mucoadhesive interpenetrating polymeric network (IPN) beads of ziprasidone HCl. These IPN beads were prepared by the combination cross-linking mechanism of ionotropic-gelation using zinc acetate and covalent gelation using glutaraldehyde. The formulation optimization of these SG–pectinate IPN beads of ziprasidone HCl was carried out by means of a 3^2-factorial design-assisted optimization process. In the optimization process, response surface methodology were employed and analyzed, where the impacts of SG amount and low methoxy pectin amount on the dependent responses like drug encapsulation and drug releasing of SG–pectinate IPN beads of ziprasidone HCl were optimized. Three-dimensional response surface graphs and two-dimensional response surface graphs showing the impacts of SG amount and low methoxy pectin amount on (a) drug encapsulation efficiency (%) and (b) cumulative drug release after 8 h (%) in simulated gastric fluid (pH 1.2) are presented in Fig. 6.8. The optimized SG–pectinate IPN beads of ziprasidone HCl exhibited superior encapsulation efficiency of

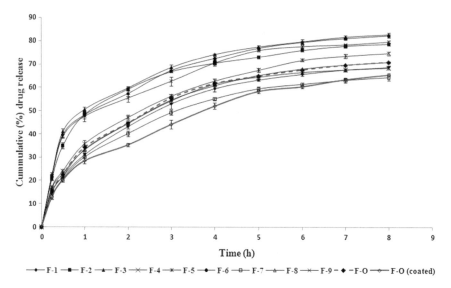

Fig. 6.6 The in vitro drug release profiles of the SG–alginate gel-coated oil-entrapped alginate floating-mucoadhesive beads of risperidone in simulated gastric fluid (pH 1.2) [Bera et al. (2015c); Copyright @ 2014, with permission from Elsevier Ltd.]

ziprasidone HCl (87.98 ± 1.15%) and average particle diameter of 2.17 ± 0.13 mm. SEM image of optimized beads demonstrated a structure of sphere-shaped having some distinctive cracks and large wrinkles onto the microparticle-surface (Fig. 6.9). In addition, in the bead surface, some polymeric derbies were seen. However, any occurrence of drug crystals on the bead surface was not evidenced by the SEM images taken. The FTIR spectroscopic results demonstrated the formation of IPN structure in between SG and low methoxy pectin. FTIR spectra studies revealed that there were no chemical interaction in between ziprasidone HCl (i.e., encapsulated drug) and SG–low methoxy pectin used to prepare IPN beads of ziprasidone HCl. In addition, DSC and XRD analyses demonstrated the physical nature as well as the stability of the drug (i.e., ziprasidone HCl) in the optimized SG–pectinate IPN beads of ziprasidone HCl. These IPN beads of ziprasidone HCl experienced sustained releasing of ziprasidone HCl during in vitro studies over 8 h in the gastric pH (1.2) (Fig. 6.10a). The sustained drug releasing may be because of decreased free volume spaces of glutaraldehyde cured IPN matrices that may limit the drug diffusion from the SG–pectinate IPN matrices. In many cases, the in vitro drug releasing followed Higuchi model kinetics with the anomalous (non-Fickian) release mechanism (Fig. 6.10b, c). The swelling behavior profiles of nearly all the SG–pectinate IPN beads of ziprasidone HCl in the gastric pH (1.2) demonstrated that the beads swelled to a smaller degree with a steady enhancement for up to 6 h (Fig. 6.11a). The optimized IPN beads similarly showed tremendous buoyancy with floating lag-time of less than 2 min and percentage buoyancy at 8 h was measured 63% (Fig. 6.12). It also presented superior mucoadhesivity with the gastric mucosa of goat (Fig. 6.12).

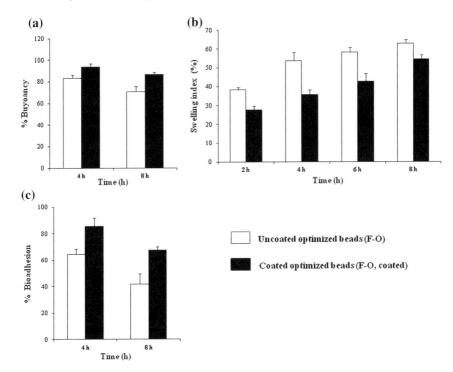

Fig. 6.7 The % buoyancy (**a**), swelling behavior (**b**) and mucoadhesivity (**c**) of the uncoated and SG–alginate gel-coated oil-entrapped alginate floating-mucoadhesive beads of risperidone in simulated gastric fluid (pH 1.2) [Bera et al. (2015c); Copyright @ 2014, with permission from Elsevier Ltd.]

6.7   SG–Alginate IPN Microparticles of Repaglinide

Kulkarni et al. (2014) prepared SG–alginate IPN microparticles of repaglinide by emulsion cross-linking and ionotropic-gelation method using a series of ionotropic cross-linker (such as aluminum chloride, barium chloride, and calcium chloride) solutions. The encapsulation efficiency of repaglinide in these SG–alginate IPN microparticles was measured from 81.10% to 91.70%. It was also detected that there was a reduction in repaglinide encapsulation efficiency with the decrease in concentration sodium alginate. The encapsulation efficiency of repaglinide of the SG–alginate microparticles prepared by aluminum ion-induced cross-linking was found as highest than the microparticles prepared by barium ion-induced cross-linking, which in turn was comparatively higher than the microparticles prepared by calcium ion-induced cross-linking. The mean size of these SG–alginate microparticles was measured as, 19.75–61.52 μm. The microparticles size was found to be more as the concentration of sodium alginate increased. The size of microparticle was found to be dependent upon the agents of ionotropic cross-linker type. The barium ion-induced SG–alginate

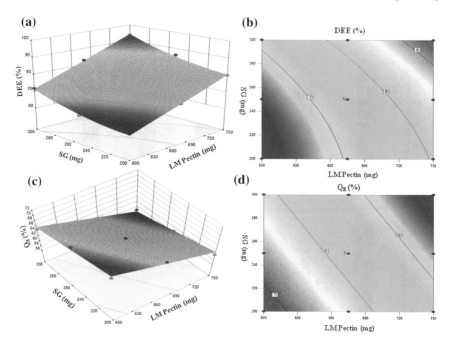

Fig. 6.8 Three-dimensional response surface graphs and two-dimensional response surface graphs showing the impacts of SG amount and low methoxy pectin amount on **a** drug encapsulation efficiency (%) and **b** cumulative drug release after 8 h (%) in simulated gastric fluid (pH 1.2) [Bera et al. (2015b); Copyright @ 2015, with permission from Elsevier Ltd.]

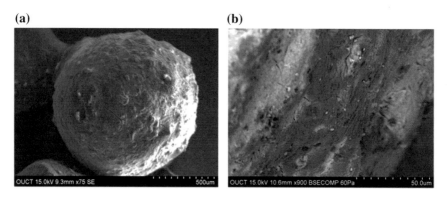

Fig. 6.9 SEM photograph of SG–pectinate floating-mucoadhesive IPN beads of ziprasidone HCl showing a rough surface with less zoom (75 X) (**a**) and presence of pores and channels with high zoom (900 X) (**b**) [Bera et al. (2015b); Copyright @ 2015, with permission from Elsevier Ltd.]

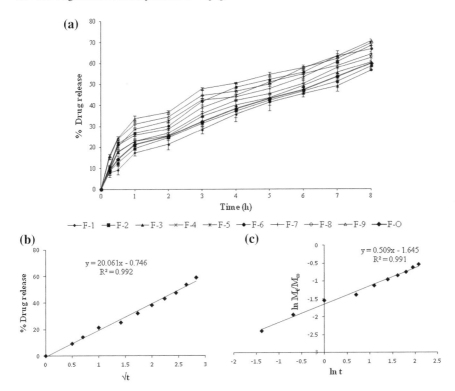

Fig. 6.10 The in vitro drug release profiles (**a**), release model (Higuchi, $Q = kt^{1/2}$) (**b**) and plot for the evaluation of diffusion exponent n and gel characteristic constant k for the drug release (**c**) from optimized SG–pectinate floating-mucoadhesive IPN beads of ziprasidone HCl [Bera et al. (2015b); Copyright @ 2015, with permission from Elsevier Ltd.]

microparticles were smaller as compared to the aluminum ion-induced SG–alginate microparticles, which in turn formed smaller microparticles than calcium ion-induced SG–alginate microparticles. The SEM analyses of SG–alginate IPN microparticles of repaglinide demonstrated smooth surface. Thermogravimetric analysis and FTIR spectroscopy confirmed the formation of IPN structure in between SG m and alginate chains in these IPN microparticles of repaglinide. XRD and DSC analyses suggested the occurrence of uniform and amorphous dispersion of repaglinide in the SG–alginate IPN matrix. The in vitro drug release from SG–alginate IPN microparticles of repaglinide was investigated in the gastric pH for initial 2 h and then, in alkaline pH for the subsequent period. The investigation showed the drug release from these IPN microparticles was sustained over 24 h. The drug release was found to be decreased from aluminum ion-induced cross-linked microparticles in comparison to that of barium ion-induced cross-linked microparticles. It was found to be lesser than that of calcium ion-induced cross-linked microparticles. The glutaraldehyde management of ionotropic cross-linked SG–alginate microparticles formed an additional decrease in drug release profile. The release of drug was found to be less as the concentra-

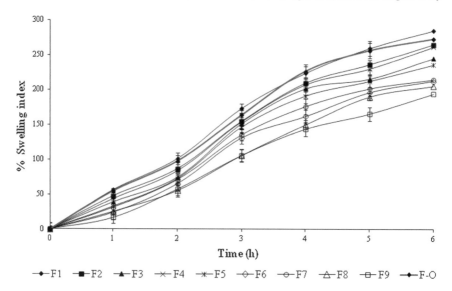

Fig. 6.11 The swelling behavior of the optimized SG–pectinate floating-mucoadhesive IPN beads of ziprasidone HCl in simulated gastric fluid (pH 1.2) [Bera et al. (2015b); Copyright @ 2015, with permission from Elsevier Ltd.]

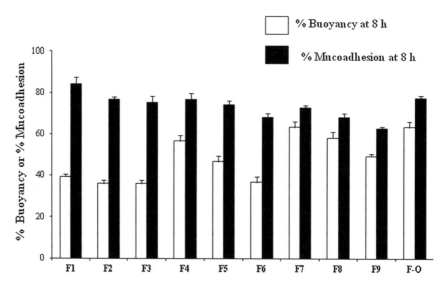

Fig. 6.12 The % buoyancy (**a**) and % mucoadhesion (**b**) of the optimized SG–pectinate floating-mucoadhesive IPN beads of ziprasidone HCl at 8 h in simulated gastric fluid (pH 1.2) [Bera et al. (2015b); Copyright @ 2015, with permission from Elsevier Ltd.]

tion of sodium alginate was enhanced in microparticles. The ionotropically cross-linked SG–alginate microparticles release the encapsulated drug rapidly; whereas the double cross-linked SG–alginate IPN microparticles produced an extended in vitro release of drug for prolonged period. The in vitro drug release from SG–alginate IPN microparticles of repaglinide followed the non-Fickian mechanism. The hypoglycemic property of SG–alginate IPN microparticles of repaglinide, in vivo, was done using an animal model, by use of streptozotocin in rats. The outcome of pristine repaglinide shows that there was a decrease in blood glucose level for up to 3 h in diabetic rats and afterward, the level of blood sugar was improved. In the case of the diabetic rats, the treatment with SG–alginate IPN microparticles of repaglinide, the percentage decrease in the level of blood glucose level was lesser in comparison to pristine repaglinide up to 3 h. However, it was augmented progressively to 81.27% up to 24 h. This showed the decreased drug release from these SG–alginate IPN microparticles of repaglinide for a prolonged period. Therefore, the developed microparticles of repaglinide were found useful to repaglinide release in a sustained manner for the effective medication of diabetes mellitus.

References

D.M.W. Anderson, W.P. Wang, The tree exudate gums permitted in foodstuffs as emulsifiers, stabilisers and thickeners. Chem. Ind. For. Prod. **14**, 73–83 (1994)

D.M.W. Anderson, Evidence of safety of gum karaya (Sterculia spp.) as a food additive. Food Addit. Contam. Part A **6**, 189–199 (1989)

K.M. Behall, D.J. Schofield, K. Lee, A.S. Powell, P.B. Mores, Mineral balance in adult men: effect of four fibers. Am. J. Clin. Nutri. **46**, 307–314 (1987)

H. Bera, S. Boddupalli, A.K. Nayak, Mucoadhesive-floating zinc-pectinate-sterculia gum interpenetrating polymer network beads encapsulating ziprasidone HCl. Carbohydr. Polym. **131**, 108–118 (2015a)

H. Bera, S.G. Kandukuri, A.K. Nayak, S. Boddupalli, Alginate-sterculia gum gel-coated oil-entrapped alginate beads for gastroretentive risperidone delivery. Carbohydr. Polym. **120**, 74–84 (2015b)

J.P. Capron, P. Zeitoun, D.A. Julien, A multicenter controlled trial of a combination of kaolin, sterculia gum, meprobamate and magnesium salts, in the irritable bowel syndrome (Authors Transl.) Gastroenterol. Clin. Biol. 5, 67–72 (1981)

D.L. Cerf, F. Irinei, G. Muller, Solution properties of gum exudates from *Sterculia urens* (karaya gum). Carbohydr. Polym. **13**, 375–386 (1990)

V.N. Deshmukh, J.K. Jadhav, D.M. Sakarkar, Formulation and *in vitro* evaluation of theophylline anhydrous bioadhesive tablets. Asian J. Pharm. **3**, 54–58 (2009)

S. Gauthami, V.R. Bhat, A monograph on gum karaya, National Institution of Nutrition, Indian Council of Medical Research, Hyderabad, India (1992)

J. Guerre, M. Neuman, Treatment of cronic colonic diseases with a new topical digestive agent, mucilage (karaya gum) combined with polyvinyl polypyrrolidone (PVPP). Med. Chirurgie Digest. **8**, 679–682 (1979)

P.R. Guru, A.K. Nayak, R.K. Sahu, Oil-entrapped sterculia gum-alginate buoyant systems of aceclofenac: Development and *in vitro* evaluation. Colloids Surf. B: Biointerf. **104**, 268–275 (2013)

E. Huttel, Treatment of acute diarrhoea in general practice. Therapeutic experiences with karaya bismuth. Die Medizinishe Welt **34**, 1383–1384 (1983)

R.V. Kulkarni, F.S. Patel, H.M. Nanjappaiah, A.A. Naikawadi, *in vitro* and *in vivo* evaluation of novel interpenetrated polymer network microparticles containing repaglinide. Int. J. Biol. Macromol. **69**, 514–522 (2014)

S. Lankalapalli, R.M. Kolapalli, Biopharmaceutical evaluation of diclofenac sodium controlled release tablets prepared from gum karaya-chitosan polyelectrolyte complexes. Drug Dev. Ind. Pharm. **38**, 815–824 (2012)

P. Meier, W.O. Seiler, H.B. Stahelin, Bulk-forming agents as laxatives in geriatric patients. Schweizerische Medizinische Wochenschrift **120**, 314–317 (1990)

D.L. Munday, J.C. Philip, Compressed xanthan gum and karaya gum matrices: hydration, erosion and drug release mechanism. Int. J. Pharm. **203**, 179–192 (2000)

A.K. Nayak, M.S. Hasnain, D. Pal, Gelled microparticles/beads of sterculia gum and tamarind gum for sustained drug release, in *Handbook of springer on polymeric gel* ed. V.K. Thakur, M.K. Thakur (Springer International Publishing, 2018) pp. 361–414

A.K. Nayak, D. Pal, Sterculia gum-based hydrogels for drug delivery applications, in *Polymeric Hydrogels as Smart Biomaterials*, ed. by S. Kalia (Springer Series on Polymer and Composite Materials, Springer International Publishing, Switzerland, 2016), pp. 105–151

C.R. Park, P.L. Munday, Evaluation of selected polysaccharide excipients in buccoadhesive tablets for sustained release of nicotine. Drug Dev. Ind. Pharm. **30**, 609–617 (2004)

D.A. Silva, A.C.F. Brito, R.C.M.D. Paula, J.P.A. Feitosa, H.C.B. Paula, Effect of mono and divalent salts on gelation of native, Na and deacetylated *Sterculia striate* and *Sterculia urens* polysaccharide gels. Carbohydr. Polym. **54**, 229–236 (2003)

B. Singh, D. Chauhan, Barium ions crosslinked alginate and sterculia gum-based gastroretentive floating drug delivery system for use in peptic ulcers. Int. J. Polym. Mater. **60**, 684–705 (2011)

B. Singh, L. Pal, Sterculia crosslinked PVA and PVA-poly(AAm)hydrogel wound dressings for slow drug delivery: mechanical, mucoadhesive, biocompatible and permeability properties. J. Mech. Behav. Biomed. Mater. **9**, 9–21 (2012)

B. Singh, N. Sharma, Modification of sterculia gum with methacrylic acid to prepare a novel drug delivery system. Int. J. Biol. Macromol. **43**, 142–150 (2008)

B. Singh, N. Sharma, Design of sterculia gum based double potential antidiarrheal drug delivery system. Colloids Surf. B: Biointerf. **82**, 325–332 (2011)

B. Singh, V. Sharma, D. Chauhan, Gastroretentive floating sterculia-alginate beads for use in antiulcer drug delivery. Chem. Eng. Res. Des. **88**, 997–1012 (2010)

B. Singh, M. Vashishtha, Development of novel hydrogels by modification of sterculia gum through radiation cross-linking polymerization for use in drug delivery. Nucl. Instr. Methods Phys. Res. Sec. B: Beam Interact. Mater Atoms. **266**, 2009–2020 (2008)

S. Singh, V. Sharma, L. Pal, Formation of sterculia polysaccharide networks by gamma rays induced graft copolymerization for biomedical applications. Carbohydr. Polym. **86**, 1371–1380 (2011)

B. Sreenivasa, R.Y. Prasanna, S. Mary, Design and studies of gum karaya matrix tablet. Int. J. Pharm. Excip. 239–242 (2000)

W. Weiping, Tragacanth and karaya, in *Handbook of hydrocolloids*, ed. by G.O. Philips, P.A. Williams (Woodhea, Cambridge, 2000), pp. 155–168

B.M. Zide, A.G. Bevin, Treatment of shallow soft tissue ulcers with an infrequent dressing technique. Anal. Plastic. Surg. **4**, 79–83 (1980)

Chapter 7
Okra Gum Based Multiple Units for Oral Drug Delivery

7.1 Okra Gum (OG)

OG is obtained from the fruits of okra plant (*Hibiscus esculentus*, family: Malvaceae) (Hirose et al. 2004). This plant is cultivated in the tropical/subtropical areas around the world. OG contains L-galacturonic acid, D-galactose as well as L-rhamnose, (Mishra et al. 2008; Sinha et al. 2015a). It is recognized as environment friendly polymeric material and inert in chemical nature. OG is water-soluble and viscous in rheological nature in the aqueous milieu (Zaharuddin et al. 2014). In the aqueous milieu, the solubility of OG is increased with the temperature increment. However, it is insoluble in a range of organic solvents like ethanol, methanol, acetone, benzene, ether, chloroform, etc. OG is biodegradable, biocompatible, and nonirritant (Sinha et al. 2015b).

7.2 Use of OG as Pharmaceutical Excipients

The encouraging solubility potential and improved rheological behavior of OG in the aqueous milieu directs the usefulness of OG as prospective polymeric excipients in numerous food, confectionary, and pharmaceutical products (Emeje et al. 2007; Ogaji 2011; Ogaji and Nnoli 2014). Since long, OG is utilized in these products as thickener, emulsifier, suspending agent, stabilizer, etc. (Kalu et al. 2007; Mishra et al. 2008). OG is also extensively studied for their potential role as the pharmaceutical excipient in the formulation of different sustained release drug delivery systems (Kalu et al. 2007; Sinha et al. 2015a, b). Their good viscous characteristic makes it suitable as the sustained drug-releasing polymeric excipient (Kalu et al. 2007; Nayak et al. 2018). OG has also been investigated as the polymeric blends along with sodium alginate

Table 7.1 Pharmaceutical applications of OG in different formulations

Formulations made of OG	Pharmaceutical applications	References
Pediatric suspensions of acetaminophen	Suspending agent	Ogaji (2011)
Paracetamol tablets	Tablet binder	Emeje et al. (2007) Ameena et al. (2010)
Paracetamol and ibuprofen tablets	Tablet binder	Patel et al. (2012)
Paracetamol matrix tablets	Matrix former, release retardant	Kalu et al. (2007)
Propranolol HCl matrix tablets	Matrix former, release retardant	Zaharuddin et al. (2014)
Film coating by OG	Film former	Ogaji and Nnoli (2014) Ogaji and Hoag (2014)
OG–alginate beads of diclofenac sodium	Encapsulating material, release retardant, matrix former	Sinha et al. (2015a)
OG–alginate composite beads of glibenclamide	Encapsulating material, release retardant, matrix former, mucoadhesive agents	Sinha et al. (2015b)

to prepare the ionotropically gelled beads for sustained releasing of various drugs (Sinha et al. 2015a, b). Pharmaceutical applications of OG in different formulations are summarized in Table 7.1.

7.3   OG–Alginate Beads of Diclofenac Sodium

In a research, by using zinc sulfate as cross-linking agent, Sinha et al. (2015a) formulated OG–alginate sustained release beads of diclofenac sodium by means of ionotropic gelation. The impacts of the ratio of sodium alginate to OG as polymer blend ratio and the zinc sulfate concentration in the ionotropic cross-linker solutions on the drug entrapment efficiency as well as cumulative in vitro drug releasing after 8 h of these OG–alginate beads of diclofenac sodium were analyzed for the formulation optimization by employing a 3^2 factorial design-based statistical optimization process. Three-dimensional response surface plots and two-dimensional contour plots describing the effects of ratio of sodium alginate to OG and zinc sulfate concentration on drug encapsulation efficiency and cumulative in vitro drug releasing after 8 h are presented in Figs. 7.1 and 7.2, respectively. The overlay plot presenting the region of optimal process variable settings is shown in Fig. 7.3. These response surface methodology analyses results depicted the augmentation in the drug encapsulation efficiency of these prepared diclofenac sodium-containing beads as the sodium alginate to OG ratio in polymeric blends decreased and concentration of zinc sulfate increased. Conversely, there was an increment in the cumulative in vitro release of encapsulated diclofenac sodium at 8 h was observed as the ratio of sodium alginate to OG as well as the concentration of zinc sulfate increased. These optimized OG–alginate beads of diclofenac sodium showed $89.27 \pm 3.58\%$ of drug encapsulation efficiency while the bead size of this optimized formulation was observed as 1.25 ± 0.12 mm. Furthermore, the cumulative in vitro drug release was found $43.73 \pm 2.83\%$ at 8 h of drug-releasing time period. With the help of scanning electron microscopy (SEM), surface morphology of optimized OG–alginate beads of diclofenac sodium was analyzed. Images of SEM study indicated that surface of the prepared beads was rough with few characteristic wrinkles, cracks, and derbies of polymers (Fig. 7.4). Fourier-transform infrared (FTIR) spectroscopy studies revealed that the compatibility of the encapsulated drug with polymers used (OG and sodium alginate) to prepare OG–alginate beads of diclofenac sodium. X-ray diffraction (XRD) results demonstrated the crystalline character of the pure drug (diclofenac sodium). In vitro drug releasing from these OG–alginate beads demonstrated prolonged in vitro release of encapsulated drug over 8 h (Fig. 7.5). The major quantity of in vitro drug releasing during initial stage of this in vitro drug release study of these formulated beads could possibly be due to the consequence of surface-adhered drug crystals, which were moreover indicated in the SEM images of the bead surface. From these formulated beads, very less amount of diclofenac sodium released up to 2 h in the acidic pH (1.2), whereas it was observed faster drug release in the alkaline pH (7.4). Formulated OG–alginate beads of diclofenac sodium demonstrated sustained release of drug over a period of 8 h and followed the controlled release pattern (i.e., zero-order kinetics) with a mechanism of super case-II transport. It was also noticed that the pH of the swelling media affects swelling and degradation of the prepared optimized OG–alginate beads of diclofenac sodium (Fig. 7.6). This kind of swelling behavior is very appropriate for intestinal drug delivery systems as these OG-based beads were observed to get swiftly dissolved in the intestinal pH.

7.4   OG–Alginate Mucoadhesive Beads of Glibenclamide

The same group, in another study, prepared OG–alginate mucoadhesive sustained release beads of glibenclamide by means of ionotropic gelation by using calcium chloride as cross-linking agent (Sinha et al. 2015b). Drug encapsulation efficiency of the formulated OG–alginate beads of glibenclamide was reported in the range of $64.19 \pm 2.02\%$ to $91.86 \pm 3.24\%$. The results of drug encapsulation efficiency indicated that the encapsulation of drug (here glibenclamide) in these formulated beads increases as the ratio of sodium alginate and OG decreased, i.e., increase in OG quantity in the polymeric blends and as the concentration of calcium chloride increased in the ionotropic cross-linking solution. This may be occurred due to high degree of ionotropic cross-linking by the calcium ions. Sizes of the beads were observed within the range of 1.12 ± 0.11 mm to 1.28 ± 0.15 mm. Image of the SEM analyses of formulated OG–alginate beads of glibenclamide showed that beads were approx-

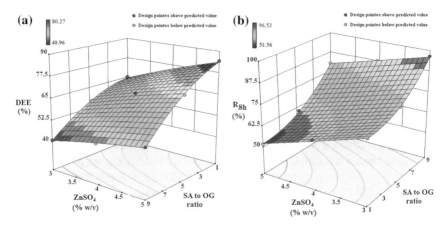

Fig. 7.1 Three-dimensional response surface plots describing the effects of ratio of sodium alginate to OG and zinc sulfate concentration on drug encapsulation efficiency and cumulative in vitro drug releasing after 8 h [Sinha et al. (2015a); Copyright @ 2015, with permission from Elsevier B.V.]

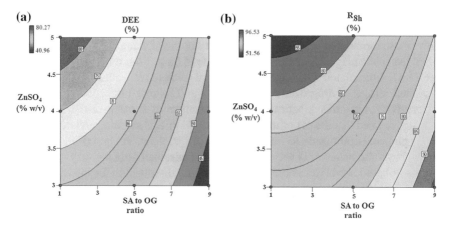

Fig. 7.2 Two-dimensional contour plots describing the effects of ratio of sodium alginate to OG and zinc sulfate concentration on drug encapsulation efficiency and cumulative in vitro drug releasing after 8 h [Sinha et al. (2015a); Copyright @ 2015, with permission from Elsevier B.V.]

imately spherical in shape and devoid of any agglomeration (Fig. 7.7). SEM images showed that bead surfaces were very irregular with few distinguished wrinkles and cracks. FTIR studies indicated that encapsulated drug preserved its characteristics after the formulation and found no drug–excipient interaction in formulated beads of OG –alginate beads of glibenclamide. In vitro drug release studies of all these for-

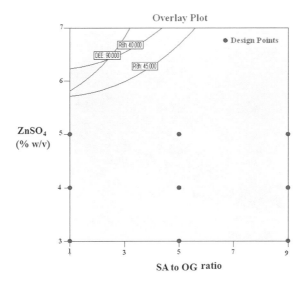

Fig. 7.3 The overlay plot indicating the region of optimal process variable settings [Sinha et al. (2015a); Copyright @ 2015, with permission from Elsevier B.V.]

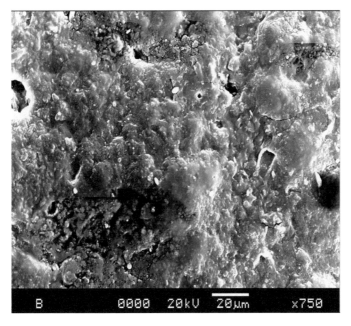

Fig. 7.4 The surface morphology of optimized OG–alginate beads of diclofenac sodium visualized by SEM [Sinha et al. (2015a); Copyright @ 2015, with permission from Elsevier B.V.]

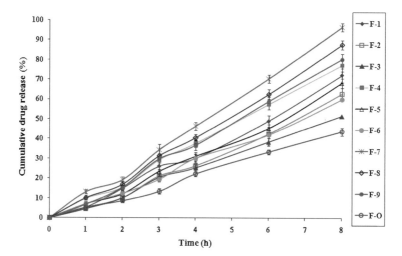

Fig. 7.5 In vitro drug release from various OG–alginate beads of diclofenac sodium [Sinha et al. (2015a); Copyright @ 2015, with permission from Elsevier B.V.]

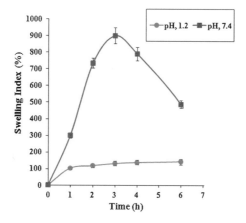

Fig. 7.6 Swelling behavior of optimized OG–alginate beads of diclofenac sodium in acidic pH (0.1 N HCl, pH 1.2) and alkaline pH (phosphate buffer, pH 7.4) [Sinha et al. (2015a); Copyright @ 2015, with permission from Elsevier B.V.]

mulated OG–alginate beads indicated sustained release of glibenclamide over 8 h, which were performed in the acidic media (pH, 1.2) for the initial 2 h and after that in the alkaline media (pH, 7.4) for 6 h (Figs. 7.8 and 7.9). These formulated beads

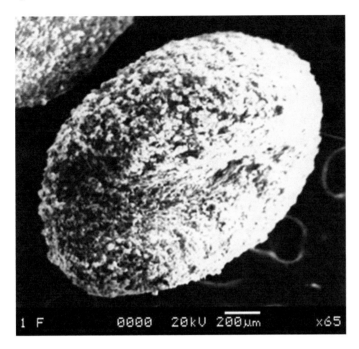

Fig. 7.7 SEM image of formulated OG–alginate beads of *glibenclamide* [Sinha et al. (2015b); Copyright @ 2014, with permission from Elsevier B.V.]

of OG–alginate beads of glibenclamide followed a controlled release pattern (i.e., zero-order kinetics) with a releasing mechanism of super case-II transport. From the overall results, formulation F-5 OG–alginate beads of glibenclamide prepared using sodium alginate to OG ratio in polymer blends of 1:1 and 8% calcium chloride concentration in cross-linking solutions was further investigated as optimized formulation. It was noticed that the pH of the swelling media affects swelling behavior of the OG–alginate beads of glibenclamide (Fig. 7.10). These OG–alginate-based beads demonstrated exceptional biomucoadhesivity in the intestinal mucosal tissue of goat (Fig. 7.11).

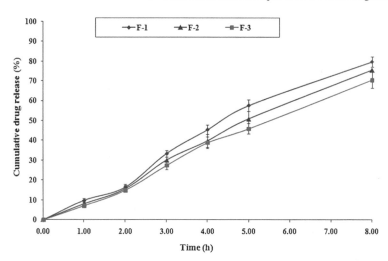

Fig. 7.8 Effect of sodium alginate to OG on in vitro drug release from F-1 [4:1], F-2 [2.5:1], and F-3 [1:1] OG–alginate beads of glibenclamide in 0.1 N HCl (pH, 1.2) for first 2 h, and then in phosphate buffer (pH, 7.4) for next 6 h (calcium chloride concentration in cross-linking solutions was 4%) [Sinha et al. (2015b); Copyright @ 2014, with permission from Elsevier B.V.]

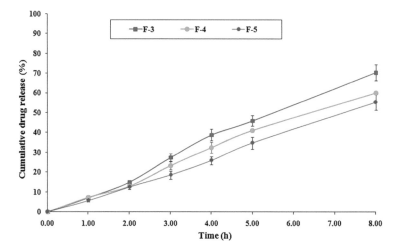

Fig. 7.9 Effect of cross-linker concentrations (calcium chloride) in cross-linking solutions on in vitro drug release from F-3 [4%]; F-4 [6%] and F-5 [8%] OG–alginate beads of glibenclamide in 0.1 N HCl (pH 1.2) for first 2 h, and then in phosphate buffer (pH 7.4) for next 6 h (sodium alginate to OG ratio in polymer blends was 1:1) [Sinha et al. (2015b); Copyright @ 2014, with permission from Elsevier B.V.]

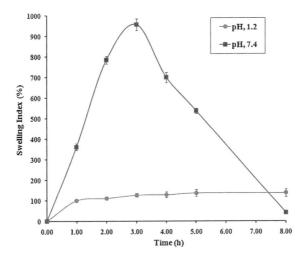

Fig. 7.10 Swelling behavior of F-5 OG–alginate beads of glibenclamide (sodium alginate to OG ratio in polymer blends was 1:1; calcium chloride concentration in cross-linking solutions was 8%) in 0.1 N HCl, pH 1.2 and phosphate buffer, pH 7.4 [Sinha et al. (2015b); Copyright @ 2014, with permission from Elsevier B.V.]

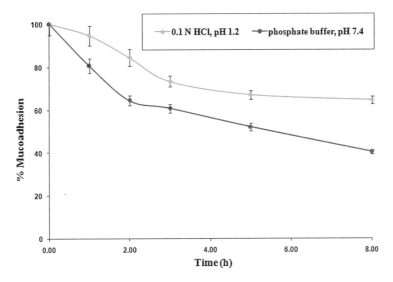

Fig. 7.11 Ex vivo wash-off behavior of F-5 OG–alginate beads of glibenclamide (sodium alginate to OG ratio in polymer blends was 1:1; calcium chloride concentration in cross-linking solutions was 8%) in 0.1 N HCl, pH 1.2 and phosphate buffer, pH 7.4 [Sinha et al. (2015b); Copyright @ 2014, with permission from Elsevier B.V.]

References

K. Ameena, C. Dilip, R. Saraswathi, P.N. Krishnan, C. Sankar, S.P. Simi, Isolation of the mucilages from *Hibiscus rosasinensis* linn. and Okra (*Abelmoschusesculentus* linn.) and studies of the binding effects of the mucilages. Asian Pac. J. Trop. Med. **3**, 539–543 (2010)

M.O. Emeje, C.Y. Isimi, O.O. Kunle, Evaluation of okra gum as a dry binder in paracetamol tablet formulations. Cont. J. Pharm. Sci. **1**, 15–22 (2007)

K. Hirose, K. Endo, K. Hasegawa, A convenient synthesis of lepidimoide from okra mucilage and its growth-promoting activity in hypocotyls. Carbohydr. Polym. **339**, 9–19 (2004)

V.D. Kalu, M.A. Odeniyi, K.T. Jaiyeoba, Matrix properties of a new plant gum in controlled drug delivery. Arch. Pharmacal Res. **30**, 884–889 (2007)

A. Mishra, J.H. Clark, S. Pal, Modification of Okra mucilage with acrylamide: synthesis, characterization and swelling behavior. Carbohydr. Polym. **72**, 608–615 (2008)

A.K. Nayak, T.J. Ara, M.S. Hasnain, N. Hoda, Okra gum-alginate composites for controlled releasing drug delivery, in *Applications of Nanocomposite materials in drug delivery*, ed. A.A.M. Inamuddin, A.A. Mohammad. volume in (Woodhead Publishing Series in Biomaterials, Elsevier Inc., 2018), pp. 761–785

I. Ogaji, Some physicochemical properties of acetaminophen pediatric suspensions formulated with okra gums obtained from different extraction process as suspending agent. Asian J. Pharm. **5**, 15–20 (2011)

I. Ogaji, O. Nnoli, Film coating potential of okra gum using paracetamol as a model drug. Asian J. Pharm. **4**, 130–134 (2014)

I.J. Ogaji, S.W. Hoag, Novel extraction and application of okra gum as a film coating agent using theophylline as a model drug. J. Adv. Pharm. Technol. Res. **5**, 70–77 (2014)

V.I. Patel, H.A. Patel, M. Jani, A. Shah, S. Kumar, J.A. Patel, Formulation and evaluation of okra fruit mucilage as a binder in paracetamol and ibuprofen tablet. Int. J. Pharm. Res. Sci. **1**, 156–161 (2012)

P. Sinha, U. Ubaidulla, M.S. Hasnain, A.K. Nayak, B. Rama, Alginate-okra gum blend beads of diclofenac sodium from aqueous template using $ZnSO_4$ as a cross-linker. Int. J. Biol. Macromol. **79**, 555–563 (2015a)

P. Sinha, U. Ubaidulla, A.K. Nayak, Okra (*Hibiscus esculentus*) gum-alginate blend mucoadhesive beads for controlled glibenclamide release. Int. J. Biol. Macromol. **72**, 1069–1075 (2015b)

N.D. Zaharuddin, MI Noordin, A. Kadivar(2014) The use of *Hibiscus esculentus* (Okra) gum in sustaining the release of propranolol hydrochloride in a solid oral dosage form. BioMed. Res. Int. 2014, Article ID 735891

Chapter 8
Fenugreek Seed Mucilage Based Multiple Units for Oral Drug Delivery

8.1 Fenugreek Seed Mucilage (FM)

Fenugreek, also called as *methi* in Hindi and Bengali, is a legume plant, whose botanical name is *Trigonella foenum-graecum* L.; it belongs to the family: Leguminosae (Naidu et al. 2011; Nayak et al. 2012). Fenugreek seeds are used as spice to flavor many foods and pickles. These are used in traditional medicine system of India and reported for hypoglycaemic (Kumar et al. 2005; Sharma et al. 1990), antioxidant (Kavirasan et al. 2007), laxative (Riad and El-Baradie 1959), gastric ulcer (Helmy 2011), and anti-inflammatory effect (Sindhu et al. 2012). Fenugreek seed husk is used as binder (Nitalikar et al. 2010) and granulating agent (Avachat et al. 2007). Due to higher mucilage content, the seeds of fenugreek when exposed to water, it swells and turns out to be slick (Petropoulos 2002). FM is prepared by boiling the fenugreek seed powder in ethanol. The treated powder of fenugreek seed is immersed in water, stirred, and then filtered out. This filtrate material is concentrated using a vacuum-type dryer set, and then mixed with 96% ethyl alcohol. After that, this concentrated filtrate is stored for overnight in a refrigerator to precipitate out the mucilage, which is separated, thoroughly washed by means of distilled water, and dried. This dried material is known as FM. FM is a galactomannan and consists of a $(1 \rightarrow 4)$ β-D-mannan skeleton to which α-D-galactopyranosyl groups are connected at the position of O-6 of D-mannopyranosyl residues with D-galactose and D-mannose ratio, 1:1 or 1:1.2 (Chang et al. 2011; Mishra et al. 2006). It also contains small amount of sugars other than mannose and galactose (Mathur and Mathur 2005).

Table 8.1 Pharmaceutical applications of FM in different formulations

Formulations made of FM	Pharmaceutical applications	References
ZnO suspension	Suspending agent	Nayak et al. (2012)
Tablets	Binder	Kulkarni et al. (2002), Nitalikar et al. (2010), Sabale et al. (2009)
Diazepam nasal gels	Gelling agent	Dutta and Bandyopadhyay (2005)
Diclofenac potassium gel	Gelling agent	Mundhe et al. (2012)
Propranolol HCl matrix tablets	Matrix former as well as release retardant	Nokhodchi et al. (2008)
Diclofenac sodium matrix tablets	Matrix former as well as release retardant	Yadhav et al. (2012)
Fast dissolving tablets amlodipine besylate	Disintegrant	Sukhavasi and Kishore (2012)
FM–alginate mucoadhesive beads of metformin HCl	Encapsulating material, release retardant, matrix former, mucoadhesive agents	Nayak et al. (2013a)
FM–pectinate mucoadhesive beads of metformin HCl	Encapsulating material, release retardant, matrix former, mucoadhesive agents	Nayak et al. (2013b)
FM–gellan gum mucoadhesive beads of metformin HCl	Encapsulating material, release retardant, matrix former, mucoadhesive agents	Nayak and Pal (2014)

8.2 Use of FM as Pharmaceutical Excipients

FM is established as pharmaceutical excipients. In pharmaceutical industry, usefulness of FM is previously recognized as binder (Kulkarni et al. 2002; Sabale et al. 2009), gelling agent [83, Gowthamarajan et al. 2002), disintegrant (Sukhavasi and Kishore 2012), matrix former [84], suspending agent [29], and release retardant [84]. It has been further studied for its potential to form the matrix and as release-retardant excipient in the preparation of diverse sustained drug-releasing systems. FM has mucoadhesive property also (Nayak and Pal 2014). The applicability of FM as potential natural mucoadhesive polymeric blends with sodium alginate, low-methoxy pectin, and gellan gum for the formulation of mucoadhesive beads is discussed. Therefore, this chapter deals with the formulation and evaluation of FM-based ionotropically gelled mucoadhesive beads for oral delivery made of FM–alginate (Nayak et al. 2013a), FM–gellan gum (Nayak and Pal 2014), and FM–pectinate blends (Nayak et al. 2013b). Pharmaceutical applications of FM in different formulations are summarized in Table 8.1.

8.3   FM–Alginate Mucoadhesive Beads of Metformin HCl

The efficacy of FM as promising natural biomucoadhesive polymeric blends with sodium alginate for the preparation of sustained drug-releasing mucoadhesive beads by means of calcium-cation-induced ionotropic gelation was assessed. Metformin HCl was studied as model drug (Nayak et al. 2013a). The impacts of sodium alginate to FM ratio (polymer blends ratio) and concentration of calcium chloride on the drug encapsulation efficiency of these FM-based beads after 10 h and cumulative in vitro drug release were optimized and analyzed by means of 3^2-factorial design. Increase in the efficiency of drug encapsulation and diminution in cumulative in vitro release of drug after 10 h were reported with the diminution of sodium alginate to FM ratio and increase in the calcium chloride concentration. Three-dimensional response surface plots and two-dimensional contour plots describing the effects of ratio of sodium alginate to FM and calcium chloride concentration on drug encapsulation efficiency and cumulative in vitro drug releasing after 10 h are presented in Figs. 8.1 and 8.2, respectively. The metformin HCl encapsulation efficiency of all the FM–alginate beads was observed in the range of $71.63 \pm 2.32\%$ to $95.08 \pm 3.73\%$. The bead sizes were reported in the range of 0.92 ± 0.05 mm to 1.30 ± 0.14 mm. Scanning electron microscopy (SEM) images of metformin HCl-loaded FM–alginate bead surface indicated exceptionally irregular surface with a few distinctive wrinkles and cracks, which may possibly be occurred due to partial collapsing of network of polymeric gel in drying process (Fig. 8.3). SEM image also indicated few polymeric derbies and crystals of drug on the surface of bead due to simultaneous gelation of the polymer blend matrix and movement toward the surface along with water in drying process, correspondingly. Results of Fourier-transform infrared (FTIR) spectroscopy analyses indicated that no interaction was found among drug and the polymers were employed as blends (sodium alginate–FM) within the prepared optimized FM–alginate beads of metformin HCl. All these prepared FM–alginate beads of metformin HCl exhibited prolonged sustained release of drug over 10 h of in vitro release studies, which was performed in 0.1 N HCl (pH, 1.2) for the initial 2 h and after that in phosphate buffer (pH, 7.4) for the leftover study period (Fig. 8.4). In vitro releasing of metformin HCl from these prepared FM–alginate beads of metformin HCl in the acidic milieu was found slower (i.e., <15.50% after 2 h) due to retrenchment of the alginate-based ionotropically gelled matrix at the acidic pH. Release of metformin HCl was found reasonably faster in the phosphate buffer, pH 7.4 due to increased rate of swelling of these prepared FM–alginate beads of metformin HCl in the alkaline pH milieu. Cumulative in vitro releasing of metformin HCl was found $69.78 \pm 2.43\%$ to $95.70 \pm 4.26\%$ after 10 h from these biopolymeric beads of metformin HCl. The in vitro release of metformin HCl over 10 h was reported to comply with the zero-order kinetic model as the best-fit model and which was also found very much close to Weibull model as well as Korsmeyer–Peppas model. A mechanism of super case-II transport was substantiated to metformin HCl releasing, which can be endorsed by the reason of polymer dissolution, enlargement of the polymeric chain, and/or relaxation of polymeric matrices. Swelling behavior of these ionotropically gelled FM–alginate beads of metformin HCl was assessed in acidic pH milieu (pH 1.2) and alkaline pH milieu (pH 7.4). Swelling index of the prepared biopolymeric beads of metformin HCl was found initially lesser in the acidic pH milieu (pH 1.2) with respect to that in the alkaline pH milieu (pH 7.4) (Figs. 8.5 and 8.6). The utmost swelling of FM–alginate beads of metformin HCl was reported at 2–3 h in the alkaline pH milieu (pH 7.4) and after that the erosion and dissolution of the polymeric matrices of ionotropically gelled FM–alginate had occurred. The ex vivo wash off for the determination of mucoadhesivity behavior of the prepared FM–alginate mucoadhesive beads of metformin HCl using intestinal mucosa of goat was done at both gastric and intestinal pHs for 8 h. The wash off was found more rapidly at intestinal pH with respect to that in the gastric pH (Figs. 8.7 and 8.8). The percentage of beads adhering to the mucosal tissue of intestine in the gastric pH was reported in the range of $56.53 \pm 2.52\%$ to $73.85 \pm 2.86\%$ over 8 h of ex vivo wash off; while this was in the intestinal pH which varied from $38.55 \pm 3.25\%$ to $48.60 \pm 0.86\%$. The enhanced mucoadhesion with absorbing surfaces of the small intestine to lengthen the gastric residence may be beneficial to deliver metformin HCl at the absorbing site over prolonged period in the controlled manner. In vivo performance of the optimized FM–alginate mucoadhesive beads of metformin HCl was tested using the alloxan-induced diabetic albino rats. The results of this in vivo study indicated a significant hypoglycemic outcome over prolonged period after the oral administration of optimized FM–alginate mucoadhesive beads of metformin HCl (Fig. 8.9).

8.4   FM–Pectinate Mucoadhesive Beads of Metformin HCl

Nayak et al. (2013a) developed mucoadhesive beads of metformin HCl, which was made of FM–low-methoxy pectin polymer blends. In this work, FM–pectinate mucoadhesive beads of metformin HCl were formulated via the ionotropic gelation by means of ionotropic cross-linking solutions of calcium chloride. For the formulation optimization, 3^2 factorial designs were employed with help of Design-Expert® Version 8.0.6.1 software. The impacts of two independable variables like of FM amounts and low-methoxy pectin amounts on the dependable responses like encapsulation efficiency of metformin HCl and cumulative metformin HCl releasing after 10 h were examined by means of the response surface methodology. On the basis of the analyses of formulation optimization, the optimal possible process variable setting employed was selected as FM amount of 698.31 mg and low-methoxy pectin amount of 348.72 mg. The augmentation of encapsulation efficiency of metformin HCl and reduction in the cumulative metformin HCl releasing after 10 h were noticed as both the polymer contents (low-methoxy pectin and FM) increased in the formulated FM–pectinate beads. Three-dimensional response surface plots and two-dimensional corresponding contour plots showing the effects of FM amounts and low-methoxy pectin amounts on encapsulation efficiency of metformin HCl and cumulative metformin HCl releasing after 10 h are presented in Fig. 8.10a–d, respectively. The obtained desirability plot indicating desirable regression ranges and the

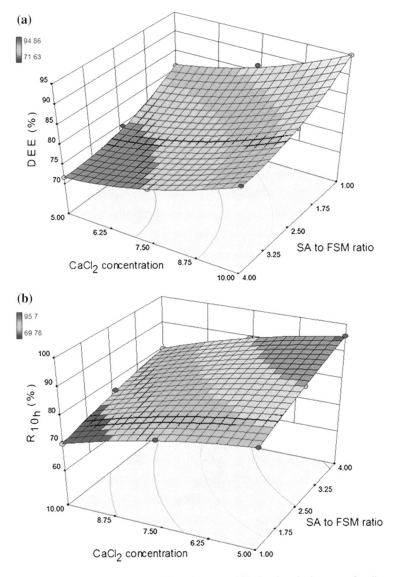

Fig. 8.1 Three-dimensional response surface plots (**a** and **b**) showing the impacts of sodium alginate to FM ratio and cross-linker concentration to prepare FM–alginate beads of metformin HCl on encapsulation efficiency of metformin HCl (DEE %) and cumulative drug release after 10 h (R_{10h} %) [Nayak et al. (2013a); Copyright @ 2012, with permission from Elsevier B.V.]

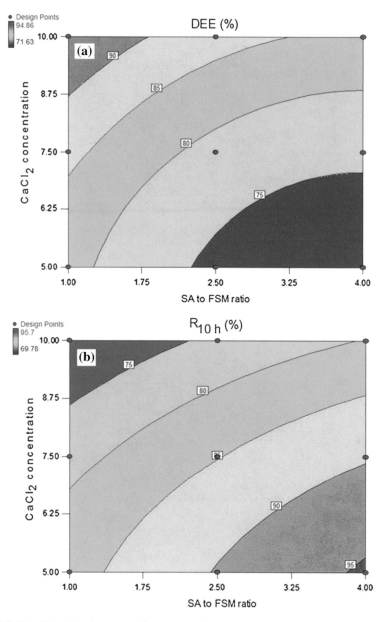

Fig. 8.2 Two-dimensional corresponding contour plots (**a** and **b**) showing the impacts of sodium alginate to FM ratio and cross-linker concentration to prepare FM–alginate beads of metformin HCl on encapsulation efficiency of metformin HCl (DEE %) and cumulative drug release after 10 h (R$_{10h}$ %) [Nayak et al. (2013a); Copyright @ 2012, with permission from Elsevier B.V.]

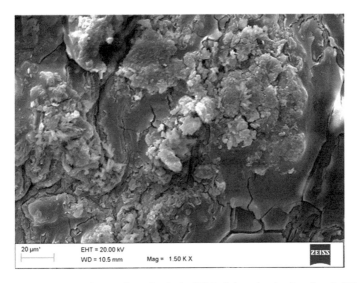

Fig. 8.3 SEM photograph of the surface of optimized FM–alginate beads of metformin HCl [Nayak et al. (2013a); Copyright @ 2012, with permission from Elsevier B.V.]

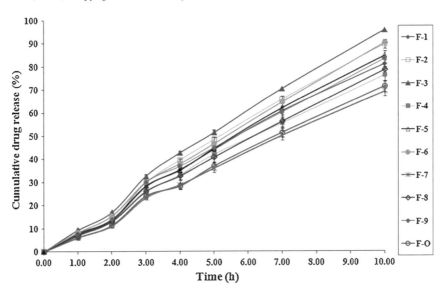

Fig. 8.4 In vitro drug release from various FM–alginate mucoadhesive beads of metformin HCl [Nayak et al. (2013a); Copyright @ 2012, with permission from Elsevier B.V.]

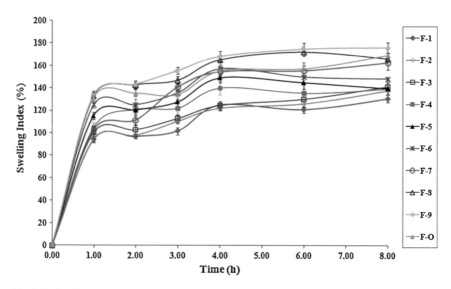

Fig. 8.5 Swelling behavior of FM–alginate beads of metformin HCl in 0.1 N HCl, pH 1.2 [Nayak et al. (2013a); Copyright @ 2012, with permission from Elsevier B.V.]

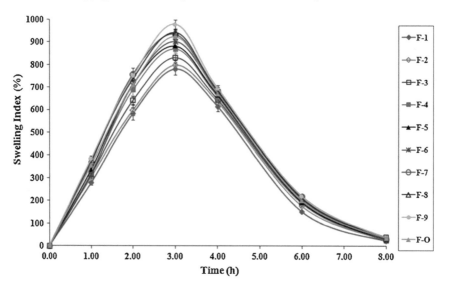

Fig. 8.6 Swelling behavior of FM–alginate beads of metformin HCl in phosphate buffer, pH 7.4 [Nayak et al. (2013a); Copyright @ 2012, with permission from Elsevier B.V.]

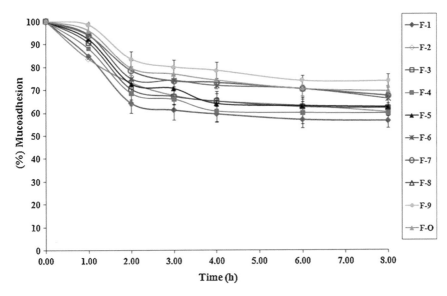

Fig. 8.7 Result of ex vivo wash-off test to assess mucoadhesive properties of FM–alginate beads of metformin HCl in 0.1 N HCl, pH 1.2 [Nayak et al. (2013a); Copyright @ 2012, with permission from Elsevier B.V.]

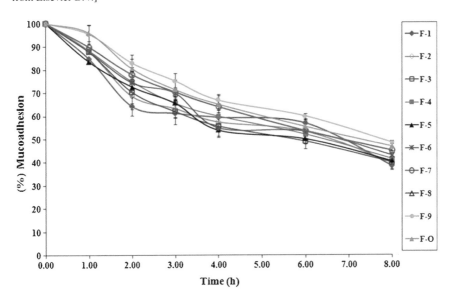

Fig. 8.8 Result of ex vivo wash-off test to assess mucoadhesive properties of FM–alginate beads of metformin HCl in phosphate buffer, pH 7.4 [Nayak et al. (2013a); Copyright @ 2012, with permission from Elsevier B.V.]

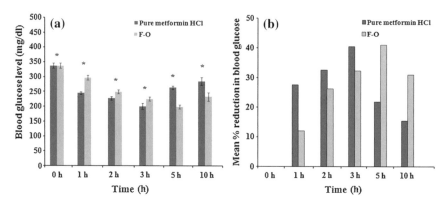

Fig. 8.9 **a** Comparative in vivo blood glucose level in alloxan-induced diabetic rats after oral administration of pure metformin HCl and FM–alginate mucoadhesive beads of metformin HCl; **b** Comparative in vivo mean percentage reduction in blood glucose level in alloxan-induced diabetic rats after oral administration of pure metformin HCl and FM–alginate mucoadhesive beads of metformin HCl [Nayak et al. (2013a); Copyright @ 2012, with permission from Elsevier B.V.]

overlay plot indicating the region of optimal process variable settings are presented in Fig. 8.10e, f, respectively. The optimized FM–pectinate beads of metformin HCl clearly demonstrated encapsulation efficiency of 96.03 ± 4.67% metformin HCl and 56.64 ± 1.47% of cumulative metformin HCl releasing after 10 h. Amount of low-methoxy pectin and FM both employed as polymeric blend extensively influenced encapsulation efficiency of metformin HCl (%) ($p < 0.05$). The drug loading and encapsulation of drug in these formulated FM–pectinate beads of metformin HCl were augmented with the increase of low-methoxy pectin and FM amounts in polymeric blends, which may perhaps be due to rise in polymeric blend solutions viscosity with rising amount of polymer addition. The mean particle diameter of dried FM–pectinate beads of metformin HCl was reported in a range of 1.47 ± 0.14 mm to 2.08 ± 0.18 mm. An increase in mean particle diameter was observed as the amounts of low-methoxy pectin and FM increased, which might be endorsed due to increase in polymeric blend solutions viscosity with the inclusion of both of these polymers in growing ratio which in turn enhanced droplet sizes for the period of adding up of polymeric blend solutions to the solutions of cross-linking agent. SEM images of optimized FM–pectinate beads of metformin HCl indicated spherical shape but discrete and rough surface (Fig. 8.11). The rough surface of these formulated beads demonstrated distinctive hefty wrinkles and cracks, and this may be due to the partial collapse of the polymeric gel network in drying process. However, little drug crystals and polymeric derbies were observed on the surface of beads. FTIR spectroscopy results confirmed the presence of various typical peaks of metformin HCl, low-methoxy pectin, and FM, which were emerged devoid of any considerable deviation or shifting of these typical peaks suggesting that there were no interaction among metformin HCl and the polymer blends employed (low-methoxy pectin and FM) in the developed optimized FM–pectinate beads of metformin HCl. The in vitro drug release profile of these FM–pectinate beads of metformin HCl showed prolonged release of metformin HCl over 10 h (Fig. 8.12). The in vitro releasing of metformin HCl from these formulated FM–pectinate beads in the acidic milieu was sluggish (i.e., <15% after 2 h) due to contraction of ionotropically gelled pectinate gel at the acidic pH milieu. The faster release of drug was reported in alkaline medium comparatively, most likely because of superior rate of swelling of the formulated beads in pH 7.4 (alkaline pH milieu). Cumulative metformin HCl releasing after 10 h was within the range 56.64 ± 1.47% to 93.63 ± 4.52%. Moreover, the releasing of drug from the formulated beads was further reported delayed as the amount of polymer amounts increased. The more hydrophilic character attained by these beads of higher polymer contents may probably bind better to the water to develop viscous gel, and this may obstruct the pores on beads surface. This might delay the release of drug from the formulated FM–pectinate beads of metformin HCl. The release of metformin HCl from the prepared beads was reported to obey zero-order model over 10 h. Moreover, Korsmeyer–Peppas model was also found to be nearer to best-fit zero-order model. Best fitting of zero-order model signified that release of drug from the prepared beads obeyed the pattern of controlled release. The super case-II transport mechanism was found to be obeyed by these beads, which could be managed by swelling and relaxation of polymer blend matrix. The swelling index of the prepared optimized FM–pectinate beads of metformin HCl was observed inferior in the acidic pH milieu (pH 1.2) as compared to that the alkaline pH milieu (pH 7.4) (Fig. 8.13a). This was taken place as a result of contraction of pectinate in the acidic pH milieu (pH 1.2). The utmost swelling of optimized FM–pectinate beads of metformin HCl was reported at 2–3 h in the alkaline pH milieu (pH 7.4) and after that erosion and dissolution of the polymeric matrices of ionotropically gelled FM–pectinate had occurred. The ex vivo wash off for the determination of mucoadhesivity behavior of the prepared FM–alginate mucoadhesive beads of metformin HCl using intestinal mucosa of goat was done at both gastric and intestinal pHs for 8 h. The wash off was found more rapidly at intestinal pH with respect to that in the gastric pH (Fig. 8.13b). The percentage of beads adhering to the mucosal tissue of intestine in the gastric pH was reported in the range of 56.53 ± 2.52% to 73.85 ± 2.86% over 8 h of ex vivo wash off, while this was in the intestinal pH varied from 38.55 ± 3.25% to 48.60 ± 0.86%. Outcome of ex vivo wash-off study of newly developed and optimized FM–pectinate beads of metformin HCl with the help of intestinal mucosa of goat in gastric pH and intestinal pH, suggesting that the ex vivo wash off of the formulated beads was faster in the intestinal pH milieu as compared to that of the gastric pH milieu. The percentage adhering of the FM–pectinate beads of metformin HCl beads to the intestinal mucosal tissue of goat ranged from 65.88 ± 4.56% in the gastric pH milieu over 8 h, while this was observed 33.35 ± 2.33% in the intestinal pH milieu. Thus, outcome of the wash-off study established that these formulated optimized FM–pectinate beads of metformin HCl had excellent mucoadhesivity. Assessment of antidiabetic activity of these newly developed optimized FM–pectinate beads of metformin HCl were carried out using the alloxan-induced diabetic albino rats. The

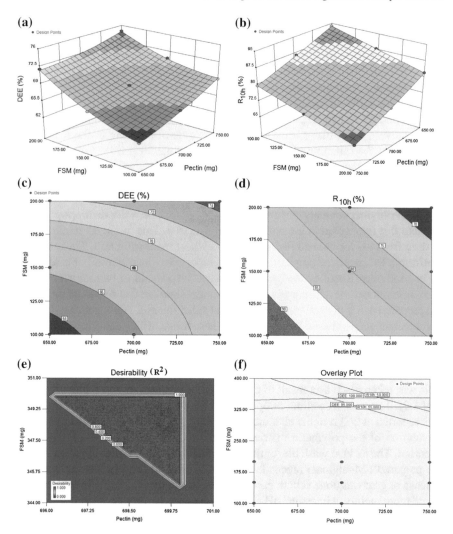

Fig. 8.10 Three-dimensional response surface plots showing the effects of amount of low-methoxy pectin (mg) and FM (mg) on **a** encapsulation efficiency of metformin HCl (DEE %) and **b** cumulative drug release after 10 h (R_{10h} %); two-dimensional corresponding contour plots showing the effects of amount of low-methoxy pectin (mg) and FM (mg) on **c** encapsulation efficiency of metformin HCl (DEE %) and **d** cumulative drug release after 10 h (R_{10h} %); the desirability plot **e** indicating desirable regression ranges and the overlay plot **f** indicating the region of optimal process variable settings [Nayak et al. (2013b); Copyright @ 2013, with permission from Elsevier Ltd.]

results of this in vivo study indicated a significant hypoglycemic outcome over 10 h after the oral administration of optimized FM–pectinate mucoadhesive beads of metformin HCl (Fig. 8.14).

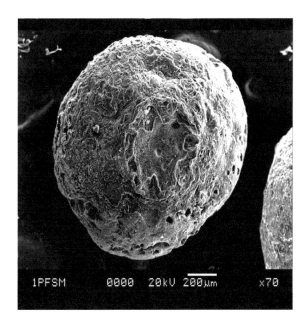

Fig. 8.11 SEM photograph of optimized FM–pectinate beads of metformin HCl [Nayak et al. (2013b); Copyright @ 2013, with permission from Elsevier Ltd.]

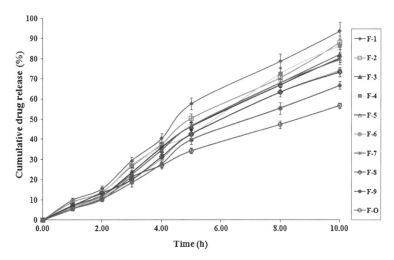

Fig. 8.12 in vitro drug release from various FM–pectinate beads of metformin HCl [Nayak et al. (2013b); Copyright @ 2013, with permission from Elsevier Ltd.]

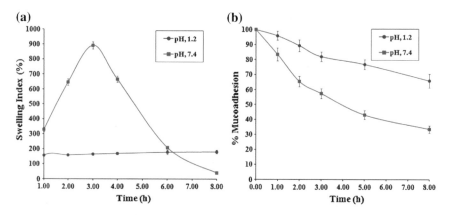

Fig. 8.13 **a** Swelling behavior of optimized FM–pectinate beads of metformin HCl in 0.1 N HCl (pH 1.2), and phosphate buffer (pH 7.4); **b** mucoadhesive behavior of optimized FM–pectinate beads of metformin HCl in 0.1 N HCl (pH 1.2), and phosphate buffer (pH 7.4) [Nayak et al. (2013b); Copyright @ 2013, with permission from Elsevier Ltd.]

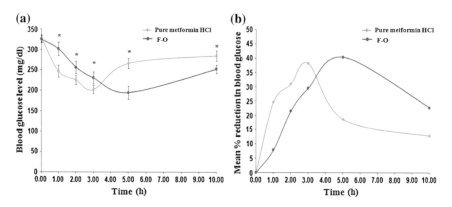

Fig. 8.14 **a** Comparative in vivo blood glucose level in alloxan-induced diabetic rats after oral administration of pure metformin HCl (standard) and optimized FM–pectinate beads of metformin HCl; **b** Comparative in vivo mean percentage reduction in blood glucose level in alloxan-induced diabetic rats after oral administration of pure metformin HCl (standard) and optimized FM–pectinate beads of metformin HCl [Nayak et al. (2013b); Copyright @ 2013, with permission from Elsevier Ltd.]

8.5 FM–Gellan Gum Mucoadhesive Beads of Metformin HCl

Nayak and Pal (2014) developed FM–gellan gum mucoadhesive beads of metformin HCl via the ionotropic gelation by means of ionotropic cross-linking solutions of calcium chloride. For the formulation optimization, a 3^2 factorial design was applied with help of Design-Expert® Version 8.0.6.1 software to explore the independent

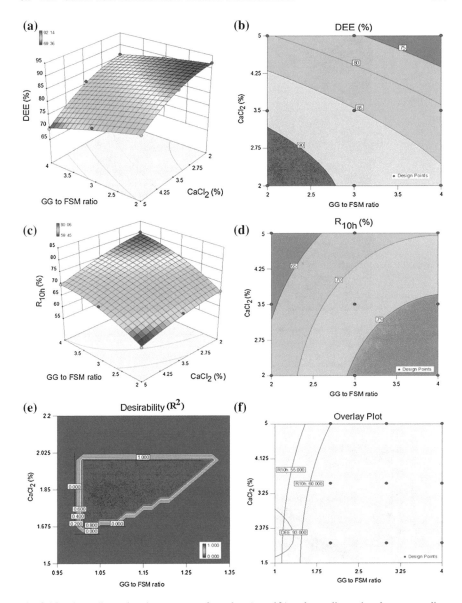

Fig. 8.15 Three-dimensional response surface plots (**a** and **b**) and two-dimensional corresponding contour plots (**c** and **d**) showing the impacts of gellan gum to FM ratio and cross-linker concentration to prepare FM–gellan gum beads of metformin HCl on encapsulation efficiency of metformin HCl (DEE %) and cumulative drug release after 10 h (R_{10h} %); the desirability plot (**e**) indicating desirable regression ranges and the overlay plot (**f**) indicating the region of optimal process variable settings [Nayak and Pal (2014); Copyright @ 2014, with permission from Elsevier Ltd.]

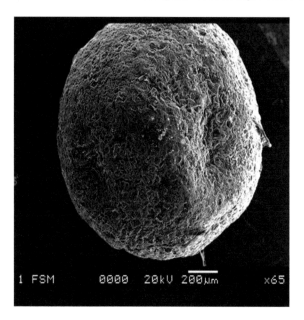

Fig. 8.16 SEM photograph of optimized FM–gellan gum beads of metformin HCl [Nayak and Pal (2014); Copyright @ 2014, with permission from Elsevier Ltd.]

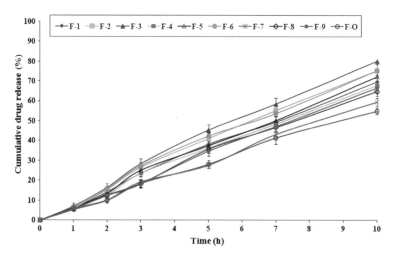

Fig. 8.17 In vitro drug release from various FM–gellan gum beads of metformin HCl [Nayak and Pal (2014); Copyright @ 2014, with permission from Elsevier Ltd.]

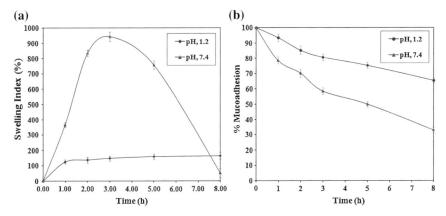

Fig. 8.18 a Swelling behavior of optimized FM–gellan gum beads of metformin HCl in 0.1 N HCl (pH 1.2), and phosphate buffer (pH 7.4); **b** Mucoadhesive behavior of optimized FM–gellan gum beads of metformin HCl in 0.1 N HCl (pH 1.2), and phosphate buffer (pH 7.4) [Nayak and Pal (2014); Copyright @ 2014, with permission from Elsevier Ltd.]

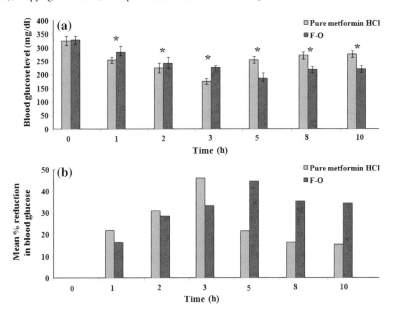

Fig. 8.19 a Comparative in vivo blood glucose level in alloxan-induced diabetic rats after oral administration of pure metformin HCl (standard) and optimized FM–gellan gum mucoadhesive beads of metformin HCl. **b** Comparative in vivo mean percentage reduction in blood glucose level in alloxan-induced diabetic rats after oral administration of pure metformin HCl (standard) and optimized FM–gellan gum mucoadhesive beads of metformin HCl [Nayak and Pal (2014); Copyright @ 2014, with permission from Elsevier Ltd.]

factors, i.e., ratio of gellan gum to FM and the concentration of calcium chloride on the encapsulation of drug and drug releasing of FM–gellan gum beads of metformin HCl. As per the 3^2 factorial design setting, nine trial formulation sets of FM–gellan gum beads of metformin HCl were prepared. By means of response surface methodology, impacts of the independent factors on the investigated responses were further elucidated. The three-dimensional response surface plots and the two-dimensional corresponding contour plots relating encapsulation efficiency of metformin HCl had signified the augmentation of encapsulation efficiency of metformin HCl with the decrease in the ratio of gellan gum to FM and concentration of calcium chloride (Fig. 8.15a, b, respectively). Though, a declining in the values of cumulative drug release after 10 h with decrease in the ratio of gellan gum to FM and concentration of calcium chloride was signified by the response surface plots and corresponding contour plots relating cumulative drug release after 10 h (Fig. 8.15c, d, respectively). The obtained desirability plot indicating desirable regression ranges and the overlay plot indicating the region of optimal process variable settings are presented in Fig. 8.15e, f, respectively. These optimized FM–gellan gum beads of metformin HCl were assessed for $92.53 \pm 3.85\%$ of metformin HCl encapsulation efficiency and $55.28 \pm 1.58\%$ cumulative drug release after 10 h. These prepared beads had mean particle sizes in the range of 1.09 ± 0.10 mm to 1.62 ± 0.22 mm. SEM images of optimized FM–gellan gum beads of metformin HCl indicated spherical shape beads with rough surface (Fig. 8.16). Bead surface exhibited distinctive large wrinkles, which may be caused by partially disintegrating the network of polymeric gel on drying process. Nevertheless, polymeric derbies and crystals of drug were observed on bead surface. FTIR spectra analyses demonstrated that the optimized FM–gellan gum beads of metformin HCl had significant characters of metformin HCl suggesting the nonexistence of any interaction among metformin HCl and polymers (i.e., FM–gellan gum) employed as blends in the bead development. The in vitro drug release study of FM–gellan gum beads of metformin HCl was performed for first 2 h in the 0.1 N HCl (pH, 1.2) and after that for next 8 h (a total of 10 h) in the phosphate buffer (pH, 7.4). This study demonstrated prolonged release of metformin HCl over 10 h (Fig. 8.17). Initially, the release of drug from these prepared beads in the acidic pH (1.2) milieu was deliberate (i.e., <16.30% after 2 h) and after that, it was reported comparatively quicker in alkaline pH (7.4) milieu. Cumulative metformin HCl releasing after 10 h was within $55.28 \pm 1.58\%$ to $80.06 \pm 3.36\%$ and reported significantly low ($p < 0.05$) with the decrease in the ratio of gellan gum to FM in polymeric blends and rising calcium chloride concentration. Because of the augmentation of hydrophilic characteristic of FM–gellan gum beads of metformin HCl formulated with the decrease in the ratio of gellan gum to FM might possibly bind well with water to develop viscous gel on the bead surface that may block the pores on these beads surface and consequently prolong release of drug occurred. Slow drug releasing from these formulated ionotropically gelled biopolymeric beads with higher calcium chloride concentration may possibly be due to free amount of matrix reduction and the obstructing movements of the solute through matrix bead at higher extent of cross-linking. A controlled releasing pattern (zero-order kinetics) with a super case-II transport mechanism controlled by the swelling was followed over 10 h indicating that the release of drug from different FM–gellan beads of metformin HCl followed controlled release pattern. The equilibrium in vitro swelling of optimized FM–gellan gum beads of metformin HCl was observed reliant on the pH of swelling media and it was observed lesser in the gastric pH milieu as compared to that in the intestinal pH milieu (Fig. 8.18). Equilibrium swelling of these formulated optimized beads of metformin HCl was observed at 3 h in intestinal pH, and then erosion and dissolution of the prepared beads had taken place. The ex vivo wash off of optimized FM–gellan gum beads of metformin HCl was found quicker in the intestinal pH milieu than in the gastric pH milieu (Fig. 8.18). The rapid ex vivo wash off reported in the alkaline pH milieu might be due to $–COO^-$ and other functional groups of matrix structure of FM–gellan gum ionization, which improved their solubility with abridged adhesive strength. The result corroborated good mucoadhesive prospective of optimized beads onto the gastrointestinal tract. The antidiabetic (hypoglycaemic) activity of optimized FM–gellan gum mucoadhesive beads of metformin HCl after administrating orally was studied using the alloxan-induced diabetic albino rats. Significant differences ($p < 0.05$) in the blood glucose level were observed after the administration of optimized FM–gellan gum mucoadhesive beads of metformin HCl and pure drug (here metformin HCl) at all time point evaluated. The optimized mucoadhesive beads showed the considerable hypoglycemic effect over 10 h after the oral administration of optimized FM–gellan gum mucoadhesive beads of metformin HCl (Fig. 8.19).

References

A. Avachat, K.N. Gujar, V.B. Kotwal, S. Patil, Isolation and evaluation of fenugreek seed husk as a granulating agent. Indian J. Pharm. Sci. **69**, 676–679 (2007)

Y.H. Chang, S.W. Cui, K.T. Roberts, P.K.W. Ng, Q. Wang, Evaluation of extrusion-modified fenugreek gum. Food Hydrocoll. **25**, 1296–1301 (2011)

R. Dutta, A.K. Bandyopadhyay, Development of a new nasal drug delivery system of diazepam with natural mucoadhesive agent from *Trigonella foenum-graecum* L. J. Sci. Ind. Res. **64**, 973–977 (2005)

K. Gowthamarajan, K.T. Giriraj, A. Muthukumar, N. Mahadevan, M.K. Samanta, B. Suresh, Evaluation of fenugreek seed mucilage as gelling agent. Int. J. Pharm. Excip. **1**, 16–19 (2002)

H.M. Helmy, Study the effect of fenugreek seeds on gastric ulcer in experimental rats. World J. Dairy Food Sci. **6**, 152–158 (2011)

S. Kavirasan, G.H. Naik, R. Gangabhagirathi, C.V. Anuradha, K.I. Priyadarsini, *in vitro* studies on antiradical and antioxidant activities of fenugreek (*Trigonella foenum-graecum*) seeds. Food Chem. **103**, 31–37 (2007)

G.T. Kulkarni, K. Gowthamarajan, Suresh B. Brahamajirao, Evaluation of binding properties of selected natural mucilages. J. Sci. Ind. Res. **61**, 529–532 (2002)

G.S. Kumar, A.K. Shetty, K. Sambaiah, P.V. Salimath, Antidiabetic property of fenugreek seed mucilage and spent turmeric in streptozotocin-induced diabetic rats. Nutr. Res. **25**, 1021–1028 (2005)

V. Mathur, N. Mathur, Fenugreek and other less known legume galactomannan polysaccharides: scope for developments. J. Sci. Ind. Res. **64**, 475–481 (2005)

A. Mishra, A. Yadhav, S. Pal, A. Singh, Biodegradable graft copolymers of fenugreek mucilage and polyacrylamide: a renewable reservoir to biomaterials. Carbohydr. Polym. **65**, 58–63 (2006)

M.R. Mundhe, R.R. Pagore, K.R. Biyani, Isolation and evaluation of *Triginella foenum-gracum* mucilage as gelling agent in diclofenac potassium gel. Int. J. Ayur. Herb. Med. **2**, 300–306 (2012)

M.M. Naidu, B.N. Shyamala, J.P. Naik, G. Sulochanamma, P. Srinivas, Chemical composition and antioxidant activity of the husk and endosperm of fenugreek seeds. LWT—Food Sci. Technol. **44**, 451–456 (2011)

A.K. Nayak, D. Pal, J. Pradhan, T. Ghorai, The potential of *Trigonella foenum-graecum* L. seed mucilage as suspending agent. Indian J. Pharm. Educ. Res. **46**, 312–317 (2012)

A.K. Nayak, D. Pal, *Trigonella foenum-graecum* L. seed mucilage-gellan mucoadhesive beads for controlled release of metformin HCl. Carbohydr. Polym. **107**, 31–40 (2014)

A.K. Nayak, D. Pal, S. Das, Calcium pectinate-fenugreek seed mucilage mucoadhesive beads for controlled delivery of metformin HCl. Carbohydr. Polym. **96**, 349–357 (2013a)

A.K. Nayak, D. Pal, J. Pradhan, M.S. Hasnain, Fenugreek seed mucilage-alginate mucoadhesive beads of metformin HCl: Design, optimization and evaluation. Int. J. Biol. Macromol. **54**, 144–154 (2013b)

M. Nitalikar, P.A. Patil, S.P. Dhole, D.M. Sakarkar, Evaluation of fenugreek seed husk as tablet binder. Int. J. Pharm. Res. Rev. **2**, 21–23 (2010)

A. Nokhodchi, A. Nazemiyeh, A. Khodaparast, T. Sorkh-Shahan, H. Valizadeh, J.L. Ford, An *in vitro* evaluation of fenugreek mucilage as a potential excipient for oral controlled-release matrix tablet. Drug Dev. Ind. Pharm. **34**, 323–329 (2008)

G.A. Petropoulos, Fenugreek: the genus Trigonella, in *Botany*, ed. by G.A. Petropoulos (Taylor and Francis, London, 2002), pp. 9–17

S. Riad, A.A. El-Baradie, Fenugreek mucilage and its relation to the reputed laxative action of this seed. Egyptian J. Chem. **2**, 163–168 (1959)

V. Sabale, V. Patel, A. Paranjapel, P Sabale, Isolation of fenugreek seed mucilage and its comparative evaluation as a binding agent with standard binder. Int. J. Pharm. Res. **1**, 56–62 (2009)

R.D. Sharma, T.C. Raghuram, N.S. Rao, Effect of fenugreek seeds on blood glucose and serum lipids in type I diabetes. Eur. J. Clin. Nutr. **44**, 301–306 (1990)

G. Sindhu, M. Ratheesh, G.L. Shyni, B. Nambisan, A. Helen, Anti-inflammatory and antioxidative effects of mucilage *Trigonella foenum-graecum* (Fenugreek) on adjuvant induced arthritic rats. Int. Immunopharmacol. **12**, 205–211 (2012)

S. Sukhavasi, V.S. Kishore, Formulation and evaluation of fast dissolving tablets of amlodipine besylate by using fenugreek seed mucilage and *Ocimum basilicum* gum. Int. Curr. Pharm. J. **1**, 243–249 (2012)

I.K. Yadhav, D.A. Jain, Design and development of *Triginella foenum-gracum* seed mucilage based matrix tablets of diclofenac sodium. World J Pharm. Res. **1**, 1170–1182 (2012)

Chapter 9
Potato Starch Based Multiple Units for Oral Drug Delivery

9.1 Potato Starch (PS)

PS is one of the natural polysaccharides obtained from the potatoes and has different applications in pharmaceutical as well as food technologies (Nazim et al. 2011; Wischmann et al. 2007). PS contains minimum or very little amount of protein or fat and very pure as well as fine in nature (BeMiller and Whistler 2009). Due to this, color of PS powder is clear white. PS contains just about 800 ppm phosphate bound to the starch, and this enhances the viscosity as well as swelling capacity (Shiotsubo 1983).

9.2 Use of PS as Pharmaceutical Excipients

PS was already studied as the pharmaceutical excipients by various research groups and found that it acts as gelling material (Nazim et al. 2011), binder, disintegrant (Bayor et al. 2013), matrix former, and release retardant (Jha and Bhattacharya 2008; Malakar et al. 2013). As a potential excipient, PS was also studied in the development of sustained drug release dosage forms (Malakar et al. 2013). Different pharmaceutical formulations based on PS are summarized in Table 9.1.

9.3 PS–Alginate Beads of Tolbutamide

Malakar et al. (2013) developed ionotropically gelled PS–alginate beads for sustained releasing of tolbutamide. These PS–alginate beads of tolbutamide were prepared through ionotropic gelation technique with the help of calcium chloride solution as

Table 9.1 Pharmaceutical applications of PS in different formulations

Formulations made of PS	Pharmaceutical applications	References
Hydrotrope PS gel as topical carrier for rofecoxib	Gelling agent	Nazim et al. (2011)
Tablets	Binder, disintegrant	Bayor et al. (2013)
PS-coated alginate microparticles containing diclofenac potassium	Encapsulating material, release retardant, matrix former	Maiti et al. (2012)
Sweet PS-blended sodium alginate microbeads of ibuprofen	Encapsulating material, release retardant, matrix former	Jha and Bhattacharya (2008)
PS-blended alginate beads for prolonged release of tolbutamide	Encapsulating material, release retardant, matrix former	Malakar et al. (2013)

the ionotropic cross-linker solution. For the formulation optimization, 3^2 factorial designs were employed with help of Design-Expert® Version 8.0.6.1 software. The impacts of two independable variables like PS amounts and sodium alginate amounts on the dependable responses like encapsulation efficiency of tolbutamide and cumulative tolbutamide releasing after 8 h were examined by means of the response surface methodology. Increase in the encapsulation efficiency of tolbutamide and reduction in cumulative in vitro release of tolbutamide after 8 h were observed with the increase in the PS amounts and sodium alginate amounts in the prepared PS–alginate beads of tolbutamide. The tolbutamide encapsulation efficiency of all these prepared PS–alginate beads of tolbutamide was observed in a range of $60.54 \pm 2.16\%$ to $85.57 \pm 3.24\%$, whereas the average sizes of the beads were calculated in a range of 1.02 ± 0.04 mm to 1.41 ± 0.07 mm. The optimized PS–alginate beads of tolbutamide showed $85.57 \pm 3.24\%$ of tolbutamide encapsulation efficiency and average bead size of 1.16 ± 0.05 mm. With the help of scanning electron microscopy (SEM) analysis, morphological analysis of these optimized PS–alginate beads of tolbutamide was performed at different magnifications and obtained SEM images at lower magnification showed that these PS–alginate beads were spherically shaped. Further detailed study of the surface of bead topography at higher magnification indicated some cracks and wrinkles on the PS–alginate bead surface. Furthermore, few pores on the surface of the bead were observed with a diameter of few micrometers. FTIR spectra studies revealed that there were no chemical interaction in between encapsulated drug (here tolbutamide) and PS–alginate used to prepare ionotropically gelled beads of tolbutamide. The in vitro releasing of encapsulated tolbutamide from various PS–alginate beads of tolbutamide was assessed in the acidic milieu (pH 1.2) for initial 2 h, and afterward in the alkaline milieu (pH 7.4) for the subsequently prolonged period. The results indicated a sustained drug-releasing pattern over 8 h. In the acidic milieu (pH 1.2) for initial 2 h, in vitro releasing of tolbutamide from these formulated beads was observed slower (i.e., <19%) due to contraction of ionotropically gelled alginate-based matrices at this low pH, and consequently quite faster release of drug was measured in the alkaline milieu (pH 7.4). The release of in vitro drugs was followed by a controlled releasing pattern (zero-order kinetics) with a case-II transport mechanism. The swelling of optimized PS–alginate beads of tolbutamide was lower in the acidic milieu (pH 1.2) than in the alkaline milieu (pH 7.4), initially due to shrinking of alginate-based hydrogels at the acidic pH environment.

9.4 PS-Coated Alginate Microparticles of Diclofenac Potassium

Maiti et al. (2012) prepared PS-coated alginate microparticles of diclofenac potassium by means of ionotropic gelation using calcium chloride solutions as an ionotropic cross-linking solution. These formulated microparticles were reported adequately stiff and free flowing in nature. SEM images of these formulated microparticles showed that these were spherical in shape, non-aggregated, and porous in nature. The in vitro diclofenac potassium releasing in phosphate buffer solution (pH 7.4) from various PS-coated alginate microparticles was assessed and the results demonstrated sustained release of drug over a time period of 10 h. The in vitro release of diclofenac potassium was furthermore characterized by means of the early bursting followed by the sluggish release of this encapsulated drug. Again, it was noticed that the smaller microparticles (in diameters) released maximum quantity of diclofenac potassium. This was also observed that concentrations of sodium alginate and PS were found to form the nonshrinkable and rigid PS-coated alginate microparticles, which again produced the slower releasing of encapsulated drug. The augmented calcium chloride concentrations in the ionotropic cross-linking solutions produced sustained releasing of encapsulated drug from these formulated PS-coated alginate microparticles of diclofenac potassium.

9.5 PS–Alginate Microbeads of Ibuprofen

In a research, Jha and Bhattacharya (2008) developed and evaluated PS–alginate microbeads for sustained releasing of ibuprofen. These microbeads were prepared by means of blendings of pre-gelatinized PS flour and sodium alginate through ionotropic gelation using calcium chloride solutions as an ionotropic cross-linking solution. The average sizes of these PS–alginate microbeads of ibuprofen ranged from 1.06 ± 0.006 mm to 1.20 ± 0.014 mm. It was observed that sizes of these microbeads were dependent on the concentrations of sodium alginate and PS, ibuprofen-loading amount, calcium chloride concentration (ionotropic cross-linking agent) stirring speed, and curing time of cross-linking. Drug encapsulation efficiency of these formulated ionotropically gelled PS–alginate microbeads of ibuprofen was reported in the range of $70.85 \pm 2.14\%$ to $93.53 \pm 1.12\%$. Increase in the drug encapsulation efficiency of these formulated beads was reported with the increase in the concentrations of polymers in the solutions of polymer blend. In vitro releasing of encapsulated drug from these PS–alginate microbeads of ibuprofen in phosphate buffer (pH 7.4) demonstrated sustained release of drug over the period of 10 h. The in vitro release of ibuprofen was furthermore characterized by means of the early bursting followed by the sluggish release of this encapsulated ibuprofen. The stability study was performed at different conditions of storage for 28 days for these formulated microbeads of ibuprofen. The results of this study showed that no significant alteration or interaction was found in the drug content of these formulated PS–alginate microbeads of ibuprofen, stored at the room temperature and 40 °C.

References

M.T. Bayor, E. Tuffour, P.S. Lambon, Evaluation of starch from new sweet potato genotypes for use as a pharmaceutical diluent, binder or disintegrant. J. Appl. Pharm. Sci. **3**, S17–S23 (2013)

J.N. BeMiller, R.L. Whistler, Potato starch: production, modifications and uses, in *Starch: chemistry and technology* (Academic Press. 3 edn), pp. 511–539 (2009)

A.K. Jha, A. Bhattacharya, Preparation and *in vitro* evaluation of sweet potato starch blended sodium alginate microbeads. Adv. Nat. Appl. Sci. **2**, 122–128 (2008)

A.K. Maiti, A.K. Dhara, A. Nanda, Preparation and evaluation of starch coated alginate microsphere of diclofenac potassium. Int. J. PharmTech Res. **4**, 630–636 (2012)

J. Malakar, A.K. Nayak, P. Jana, D. Pal, Potato starch-blended alginate beads for prolonged release of tolbutamide: development by statistical optimization and *in vitro* characterization. Asian J. Pharm. **7**, 43–51 (2013)

S. Nazim, M.H.G. Dehghan, S. Shaikh, A. Shaikh, Studies on hydrotrope potato starch gel as topical carrier for rofecoxib. Der. Pharm. Sinica **2**, 227–235 (2011)

T. Shiotsubo, Starch gelatinization at different temperatures as measured by enzymic digestion method. Agri. Biol. Chem. **47**, 2421–2425 (1983)

B. Wischmann, T. Ahmt, O. Bandsholm, A. Blennow, N. Young, L. Jeppesen, L. Thomsen, Testing properties of potato starch from different scales of isolations- a ringtest. J. Food Eng. **79**, 970–978 (2007)

Chapter 10
Linseed Polysaccharide Based Multiple Units for Oral Drug Delivery

10.1 Linseed Polysaccharide (LP)

LP is extracted from mature and ripe linseeds (*Linum usitatissimum* L., family: Liliaceae) (Khadse et al. 2010). Linseeds are also known as flax seeds. The mature and ripe linseeds principally contain polysaccharides of 8–10% by weight, approximately. On the acid-catalyzed hydrolysis, LP produces D-xylose, L-arabinose, L-galactose, L-rhamnose, D-glucose, and D-galacturonic acid (Nerkar and Gattani 2011). It is a water-soluble polysaccharide and is also able to produce viscous solution in the aqueous medium. It is biocompatible and biodegradable in nature (Mahant et al. 2011; Nerkar and Gattani 2011). During past few decades, it has gained technological as well as industrial significance as a potential plant polysaccharidic substance due to its biocompatibility and biodegradability. Since the ancient ages, LP is also known as a potential neutraceutical due to its role in health care (Mitra and Bhattacharya 2009). The importance of LP in controlling diabetes is already established (Mitra and Bhattacharya 2009; Mahant et al. 2011; Nerkar and Gattani 2011). It is also exploited in food and beverage materials (Mitra and Bhattacharya 2009).

10.2 Use of LP as Pharmaceutical Excipients

LP is widely investigated in the formulations of cosmeceuticals, pharmaceuticals, etc. (Khadse et al. 2010; Mahant et al. 2011). According to the previous literature, investigations were performed on the utilization of LP as suspending agent (Khadse et al. 2010), tableting excipients (Mahant et al. 2011), matrix-forming agents (Nerkar and Gattani 2011), and gelling agent (Basu et al. 2007). LP has already been studied in the formulations of oral, buccal, and nasal drug deliveries (Basu et al. 2007; Mahant et al. 2011; Nerkar and Gattani 2011). In addition, LP was already investigated

Table 10.1 Pharmaceutical applications of LP in different drug delivery formulations

Formulations made of LP	Pharmaceutical applications	References
Suspensions	Suspending agent	Khadse et al. (2010)
Tablets	Tablet binder	Wankhede and Shrivastava (2013)
Fast disintegrating tablets of glibenclamide	Disintegrant	Shirsand et al. (2013)
Mucoadhesive tablets of ranitidine HCl	Mucoadhesive agent	Mahant et al. (2011)
Mucoadhesive nasal gel of midazolam	Mucoadhesive agent, gelling agent	Basu et al. (2007)
LP–alginate mucoadhesive beads for controlled drug releasing	Encapsulating material, release retardant, matrix former	Hasnain et al. (2018)
Buccal mucoadhesive microspheres of venlafaxine	Mucoadhesive agent, matrix former	Nerkar and Gattani (2011)

as mucoadhesive agent in the development of mucoadhesive tablets (Mahant et al. 2011), mucoadhesive buccal formulations (Nerkar and Gattani 2011), and mucoadhesive nasal gels (Basu et al. 2007) by reason of its biomucoadhesive nature. Pharmaceutical applications of LP in different drug delivery formulations are summarized in Table 10.1.

10.3 LP–Alginate Mucoadhesive Beads of Diclofenac Sodium

The efficacy of LP as pharmaceutical excipient in the preparation of ionotropically cross-linking gelled LP–alginate mucoadhesive beads for the controlling release of diclofenac sodium was investigated (Hasnain et al. 2018). LP–alginate beads loaded with diclofenac sodium were formulated via ionotropically cross-linking gelation method using calcium chloride as ionotropic cross-linker agent. These ionotropically cross-linked gelled beads showed diclofenac sodium encapsulation efficiencies in these newly prepared beads which were $60.78 \pm 2.47\%$ to $93.16 \pm 4.08\%$. The average bead sizes of these ionotropically cross-linked gelled LP–alginate beads loaded with diclofenac sodium were calculated via optical microscopic technique and this was within the range from 1.17 ± 0.10 mm to 1.33 ± 0.12 mm. The larger sized beads were formed when increased concentrations of LP were employed for the preparation. Roughly, spherical-shaped beads devoid of forming any types of agglomeration with the denser and thicker coat of excipient polymers were observed in the scanning electron microscopy (SEM) photograph (Fig. 10.1). In vitro diclofenac sodium releasing from different formulated ionotropically cross-linked gelled LP–alginate

Fig. 10.1 SEM photograph of LP–alginate beads of diclofenac sodium [Hasnain et al. (2018); Copyright @ 2018, with permission from Elsevier B.V.]

beads was tested in the acidic pH (1.2) milieu for first 2 h, and afterward in the alkaline pH (7.4) milieu for the remaining period. A sustained in vitro drug releasing was observed over 8 h (Fig. 10.2). Within the initial 2 h of study in the acidic pH (1.2) milieu, a very slower in vitro drug releasing was noticed for all these tested beads. All LP–alginate beads loaded with diclofenac sodium demonstrated a prolonged sustained drug-releasing profile over 8 h with a zero-order model of drug releasing (controlled drug-releasing pattern). Among these, LP–alginate bead formulation F-5 (loaded with diclofenac sodium; formulated using LP and sodium alginate as excipient polymers in the ratio of 1:1 and calcium chloride concentration of 8%) was chosen as the optimized formulation as per the higher values of diclofenac sodium encapsulation efficiency (93.16 ± 4.08%) and slower sustained in vitro drug releasing over 8 h. The optimized LP–alginate beads loaded with diclofenac sodium displayed a pH-responsive swelling (Fig. 10.3) and excellent biomucoadhesivity prospective with the intestinal mucosal tissue (Fig. 10.4) in both the acidic and alkaline pH (pH 1.2 and 7.4, respectively). Thus, the efficacy of LP was proved as a prospective mucoadhesive and sustained drug-releasing natural polymeric agent (which is also plant derived) with sodium alginate to prepare ionotropically cross-linked gelled LP–alginate mucoadhesive beads for oral use via ionotropically cross-linking gelation method. These LP–alginate mucoadhesive and sustained drug-releasing biopolymeric beads can be formulated to encapsulate other drugs, which need sustained releasing for good bioavailability as well as higher patient compliances with the reduction of multiple dosing.

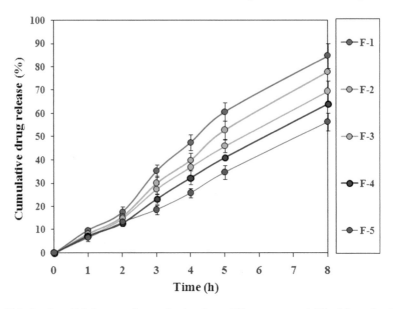

Fig. 10.2 In vitro diclofenac sodium releasing from different prepared LP–alginate beads of diclofenac sodium [Hasnain et al. (2018); Copyright @ 2018, with permission from Elsevier B.V.]

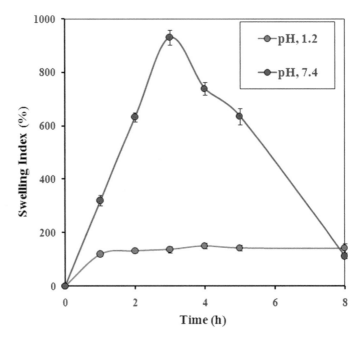

Fig. 10.3 Swelling behavior of optimized LP–alginate beads of diclofenac sodium in pH 1.2 and pH 7.4 [Hasnain et al. (2018); Copyright @ 2018, with permission from Elsevier B.V.]

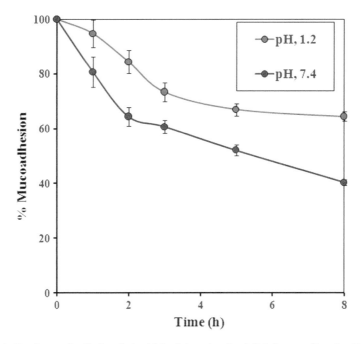

Fig. 10.4 Ex vivo wash off of optimized LP–alginate beads of diclofenac sodium in pH 1.2 and pH 7.4 [Hasnain et al. (2018); Copyright @ 2018, with permission from Elsevier B.V.]

References

S. Basu, S. Chakraborty, A.K. Bandyopadhyay, Development and evaluation of a mucoadhesive nasal gel of midazolam prepared with *Linum usitatissimum* L. seed mucilage. Sci. Pharm. **77**, 899–910 (2007)

M.S. Hasnain, P. Rishishwar, S. Rishishwar, S. Ali, A.K. Nayak, Isolation and characterization of *Linum usitatisimum* polysaccharide to prepare mucoadhesive beads of diclofenac sodium. Int. J. Biol. Macromol. **116**, 162–172 (2018)

C.D. Khadse, R.B. Kakade, V.K. Deshmukh, D.S. Mohale, Evaluation of suspending properties of mucilage of *Linnum usitatisimum* Linn. seeds. Res. J. Pharmacog. Phytochem. **2**, 208–210 (2010)

S. Mahant, N. Khurana, S. Dua, P. Thakur, I. Bakshi, Formulation and evaluation of mucoadhesive tablets using flax seed mucilage. J. Pharm. Biomed. Sci. **6**, 1–7 (2011)

A. Mitra, D. Bhattacharya, Role of flax and flax gum in health and diabetes. Indian J. Pract. Doctor. **5**, 19–22 (2009)

P.P. Nerkar, S. Gattani, *In vivo, in vitro* evaluation of linseed mucilage based buccal mucoadhesive microspheres of venlafaxine. Drug. Deliv. **18**, 111–121 (2011)

S.B. Shirsand, H. Rohini, S. Prathap, P.V. Prakash, *Linum unitatissimum* and *Lepidium sativum* mucilage in the design of fast disintegrating tablets. RGUHS J. Pharm. Sci. **3**, 19–25 (2013)

V.K. Wankhede, R.S. Shrivastava, Extraction and characterization of mucilage obtained from *Linum usitatissimum* and its use as binder in tablet formulation. Int. J. Pharm. Sci. Res. Des. **5**, 101–107 (2013)

Chapter 11
Some Other Plant Polysaccharide Based Multiple Units for Oral Drug Delivery

11.1 Tapioca Starch (TS)–Alginate Gastroretentive Mucoadhesive Floating Beads of Metoprolol Tartrate

TS is a naturally occurring starch and it is obtained from the *Manihot esculenta* root extract (BeMiller and Whistler 2009; Biswas and Sahoo 2016). The residual matter level of TS is lesser in comparison with other starch-derived plants. The content of amylase in TS is less (around, 17%) as compared to other starch-derived plants (Casas et al. 2010). These features of TS make it as a potential material in several industrial applications (BeMiller and Whistler 2009). Recently, TS is being exploited as pharmaceutical excipients such as matrix former, disintegrant, mucoadhesive agent, and release retardant in several pharmaceutical dosage forms (Casas et al. 2010; Biswas and Sahoo 2016).

Biswas and Sahoo (2016) have formed ionotropically gelled TS–alginate beads for gastroretentive delivery of metoprolol tartrate (a widely used β_1 blocker in hypertension treatment; possessing short plasma half-life of approximately 3.5 h) via the dual mechanism of mucoadhesion floatation. These TS–alginate beads of metoprolol tartrate were formed by means of ionotropic-gelation technique using calcium chloride as an ionic cross-linker and with/without integration of calcium carbonate as a gas-forming floating agent. The encapsulation efficiency of drug of these TS–alginate beads was found to be improved by more than 85% after the incorporation of TS in the ionotropically gelled alginate matrices with/without inclusion of calcium carbonate. The increasing incorporation of TS in the bead formula of TS–alginate beads of metoprolol tartrate showed better encapsulation efficiency of drug, which may be because of prevention of the leaching of drug (here metoprolol tartrate) before and after the ionotropic gelation of the TS–alginate matrices with the use of increasing incorporation of TS in the bead formula. The incorporation of calcium carbonate as gas-forming floating agent in these TS–alginate bead matrices

A. K. Nayak and M. S. Hasnain, *Plant Polysaccharides-Based Multiple-Unit Systems for Oral Drug Delivery*, SpringerBriefs in Applied Sciences and Technology, https://doi.org/10.1007/978-981-10-6784-6_11

had not found to influence the encapsulation efficiency of drug. TS–alginate beads of metoprolol tartrate formed without the incorporation of calcium carbonate showed mean particle size that ranges from 1033 ± 76.54 μm to 1288 ± 45.46 μm, whereas 1192 ± 68.28 μm to 1254 ± 88.22 μm of mean particle size range was measured in case of beads formed with the incorporation of calcium carbonate as floating agent. In the analyses of particle size, it was seen that mean particle sizes of TS–alginate beads of metoprolol tartrate was found as the TS amounts were increased in the bead formula. Further coating of HPMC K4 M onto these beads produced larger size with a particle size ranging from 1324 μm to 1410 μm. The morphology of the TS–alginate beads of metoprolol tartrate was examined through scanning electron microscopy (SEM). The photograph by SEM showed the morphological differences between TS–alginate beads as well as alginate beads. TS–alginate beads of metoprolol tartrate contained calcium carbonate (the gas-forming floating agent) which showed some little pores due to the carbon dioxide release in the gastric medium (pH 1.2). TS–alginate beads of metoprolol tartrate were characterized by the use of Fourier-transform infrared spectroscopy (FTIR) analysis and the results of the FTIR analysis showed the absence of any chemical interactions in between metoprolol tartrate and excipient polymers (sodium alginate and TS) used in the preparation of these TS–alginate beads. The gastric residence ability of TS–alginate beads of metoprolol tartrate in the gastric pH (1.2) media was evaluated by in vitro wash-off test for mucoadhesion as well as buoyancy test for floatation. The in vitro wash-off test was carried out using goat intestinal mucosa as mucosal membrane, and 28.40% to 62.20% of mucoadhesion after 4 h was considered for the TS–alginate beads of metoprolol tartrate formed without incorporating calcium carbonate. The TS–alginate bead matrices most likely turn into more rigid with the enhancement of concentration of TS in the TS–alginate bead formula, which resulted in detachment of these TS–alginate beads from goat intestinal mucosa before the swelling and adhering to the mucus gel layer. The TS–alginate beads of metoprolol tartrate formed with incorporation of calcium carbonate showed lower mucoadhesion after 2 h. In these beads, bubbles of carbon dioxide gas liberated through the reaction in between the gastric media and calcium carbonate facilitated in detachment of the TS–alginate beads from the mucus layers. Rise in the concentration of calcium carbonate in these TS–alginate bead formulas produced more liberation of carbon dioxide gas, which increased the detachment of these beads from the intestinal mucosa of goat. HPMC K4 M-coated TS–alginate beads of metoprolol tartrate exhibited more than 40% of the mucoadhesion after a period of 4 h. The stronger gel-forming capability of HPMC K4 M and slower permeation of liberated carbon dioxide gas from the gel layer may have contributed to higher degree of mucoadhesion with the intestinal mucosa of goat. The mucoadhesion kinetics was evaluated using the results of in vitro wash-off test for mucoadhesion and it was found to follow first-order kinetic model. The TS–alginate beads of metoprolol tartrate containing calcium carbonate remained buoyant for a period of 12 h by the liberating effervescence of carbon dioxide gas when prepared. The buoyancy of the TS–alginate beads contained calcium carbonate which was seen to be proportional with the calcium carbonate concentration of in bead formula. About 1.50% to 2.50% w/w of calcium carbonate incorporated TS–alginate beads of metoprolol tartrate was seen to hold up the sinking of maximum number of beads for a period of 16 h in the gastric pH. On the other hand, TS–alginate beads of metoprolol tartrate formed without calcium carbonate incorporation showed quick immersion in the gastric pH. HPMC K4 M-coated TS–alginate beads of metoprolol tartrate showed good buoyant behavior. TS–alginate beads of metoprolol tartrate coated using 8% to 12% w/w HPMC K4 M exhibited 65% of buoyancy for 12 h. The in vitro release of metoprolol tartrate from TS–alginate beads were evaluated in simulated gastric fluid solution of pH 1.2. The results of in vitro release of drug established a biphasic releasing profile of 55% to 60% of metoprolol tartrate for a period of the first hour, and subsequently a more gradual metoprolol tartrate release profile reaching 90% metoprolol tartrate after 3 to 4 h of drug release study was noticed. The incorporation of TS in these TS–alginate bead formulas may have customized the in vitro release of drug pattern by protecting the erosion as well as disintegration of the TS–alginate matrix. A significant ($p < 0.05$) decrease in in vitro release of drug pattern from these TS–alginate beads having metoprolol tartrate was noticed with enhanced amount of TS. The TS–alginate floating mucoadhesive beads of metoprolol tartrate formed with incorporation of calcium carbonate showed in vitro metoprolol tartrate release of 90% within a period of 75 min to 90 min simulated gastric fluid solution of pH 1.2. The in vitro drug release rate of these TS–alginate floating mucoadhesive beads was found to be enhanced with the increase in amount of calcium carbonate. HPMC K4 M-coated TS–alginate floating mucoadhesive beads showed a sustained metoprolol tartrate release for a period of 8 to 10 h at the coating weights of 12% w/w. The coating weights of 12% w/w HPMC K4 M onto TS–alginate floating mucoadhesive beads of metoprolol tartrate were found inadequate to decrease the in vitro drug release significantly ($p < 0.05$). The coating weight of 16% w/w HPMC K4 M onto these floating mucoadhesive beads of metoprolol tartrate produced an incomplete release of 87% and extended up to 14 h of release study. The in vitro metoprolol tartrate release from TS–alginate floating mucoadhesive beads is explained by the first-order kinetic model and an anomalous behavior (non-Fickian mechanism) indicating both diffusion of drug through the gel layer and macromolecular relaxation of the TS–alginate matrices. The pharmacokinetic and pharmacodynamic profiles of the optimized TS–alginate floating mucoadhesive beads of metoprolol tartrate were carried out in the animal model using adult healthy male New Zealand rabbits. Pharmacokinetic profile of the optimized floating mucoadhesive TS–alginate beads of metoprolol tartrate in the rabbit model showed an enhanced relative oral bioavailability of metoprolol tartrate by 87% after oral administration as compared to that of sustained release (67%) and oral solution (51%). Optimized TS–alginate floating mucoadhesive beads of metoprolol tartrate established a higher percent inhibition of isoprenaline-induced heart rate in rabbits for almost 12 h.

11.2   *Aloe Vera* Gel–Alginate Non-floating and Floating Beads of Ranitidine HCl

Aloe vera gel is obtained from leaves of parenchyma cells of *Aloe vera*. It is transparent mucilaginous jelly (Femenia et al. 1999). It contains natural polysaccharide consisted of (1, 4)-linked highly acetylated polydispersed mannans (acemannan) (Singh et al. 2012). It has the potential to heal up gastric ulcer or to protect the occurrences of gastric ulcers and these abilities have been attributed to several possible mechanisms such as healing property, anti-inflammatory activity, regulation of gastric secretion, mucous stimulatory property, *etc* (Femenia et al. 1999; Singh et al. 2012). Besides these possible remedial properties, *Aloe vera* gel polysaccharide has revealed excellent ability to be used as excipients in pharmaceuticals in the formulation of several dosage forms like sustained drug release dosage forms (Jani et al. 2007 ; Avachat et al. 2011).

In the perspective of therapeutic importance and sustained release of drug, *Aloe vera* gel was blended with sodium alginate to formulate *Aloe vera* gel-blended alginate non-floating and floating beads through ionotropic gelation using three different ionotropic cross-linkers, namely, calcium chloride, barium chloride, and aluminum chloride (Singh et al. 2012). A H_2-receptor agonist drug, ranitidine HCl, was broadly used in the treatment of ulcer and was used as model drug in this study. Calcium carbonate was used as floatation imparting material to formulate *Aloe vera* gel–alginate floating beads of ranitidine HCl. The diameter of floating beads of ranitidine HCl was larger in comparison with the diameter of non-floating beads having ranitidine HCl. By the use of SEM and FTIR, these *Aloe vera* gel–alginate non-floating and floating beads were characterized. SEM results showed that spherical-shaped beads were having rough surface in case of non-floating beads, while smooth surface of floating beads was seen in formulation using the addition of calcium carbonate. In spite of enhanced gelation by the calcium ions, some characteristic cracks were observed on the surface of floating bead that may be due to the effervescence effect by the evolved carbon dioxide before the walls get adequately rigid. FTIR analysis suggested the absence of any kinds of interactions between the ranitidine HCl and the excipients employed. The in vitro release of ranitidine HCl from *Aloe vera* gel-alginate floating and non-floating beads of ranitidine HCl was carried out in the release media of pH 2.2 buffer and in distilled water. The in vitro ranitidine HCl release results signified that the ranitidine HCl releasing from both the floating and non-floating beads has been observed less in the distilled water in comparison with that in pH 2.2 buffer in each case of different ionotropic cross-linker investigated. In addition, the in vitro ranitidine HCl release from these *Aloe vera* gel-alginate beads followed the Fickian diffusional mechanism.

11.3 Assam Bora Rice (ABR) Starch–Alginate Microbeads of Metformin HCl

ABR is accepted as festival food in Assam. From ABR, the obtained starch is sticky and waxy in nature. Sachan and Bhattyacharya (2006, 2009) evaluated the probability of ABR starch as possible natural biomucoadhesive polymer blends with sodium alginate for supporting the release of drug that is water-soluble. They developed metformin HCl-loaded ABR starch-blended alginate mucoadhesive microbeads by the micro-orifice ionotropic-gelation technique in all aqueous system using pre-gelatinized ABR blends with sodium alginate in various ratios (Sachan and Bhattyacharya 2009). The mean sizes of these formulated metformin HCl-loaded ABR starch–alginate mucoadhesive microbeads were found ranging from 0.726 ± 0.008 mm to 1.16 ± 0.009 mm. These microbeads showed good mucoadhesion in vitro wash-off test as compared with prepared microbeads by the use of non-mucoadhesive material. The in vitro release of drug from these metformin HCl-loaded ABR starch–alginate mucoadhesive microbeads was found persistent for about 12 h. The release of drug behavior and extent of its release duration were dependent on the gel strength, precision device, drug loading and polymer concentration, stirring speed, polymer blends ratio, curing time, cross-linker concentration, *etc.* Thus, the results of this study showed the application of ABR starch as polymer blends with sodium alginate to formulate microbeads for drug delivery of sustained release formulations.

11.4 Gum Cordial–Gellan Gum Beads of Metformin HCl

Gum cordial is a natural gum which is anionic in nature and is obtained from *Cordia oblique* Wild fruits. It belongs to the family of Boraginaceae. It is one such agent which has higher potential application pharmaceuticals. According to Ahuja et al. (2010), the ionotropically gelled beads made of gum cordial–gellan gum blends for sustained release of metformin. The preparation of metformin-loaded gum cordial–gellan gum beads was optimized using response surface methodology. Effects of different formulation and process variables like concentrations of gellan gum, drug (metformin HCl), gum cordial and calcium chloride, temperature, hardening time, pH, and height on entrapment of drug and its release were screened using Plackett–Burman design. The results showed that concentration of chloride and drug, temperature, and hardening time have significant effects. In addition, gum cordial–gellan gum beads batches were synthesized utilizing domain. The optimized gum cordial–gellan gum bead was found to attain the goals of adequate entrapment of drug by release of 80% in 24 h.

References

M. Ahuja, M. Yadav, S. Kumar, Application of response surface methodology to formulation of ionotropically gelled gum cordial/gellan beads. Carbohydr. Polym. **80**, 161–167 (2010)

A.M. Avachat, R.R. Dash, S.N. Shrotriya, Recent investigations of plant based natural gums, mucilages and resins in novel drug delivery systems. Indian J. Pharm. Educ. Res. **45**, 86–99 (2011)

J.N. BeMiller, R.L. Whistler, Potato starch: production, modifications and uses, in: *Starch: chemistry and technology* (Academic Press. 3rd edn., 2009) pp. 511–539

N. Biswas, R.K. Sahoo, Tapioca starch blended alginate mucoadhesive-floating beads for intragastric delivery of metoprolol tartrate. Int. J. Biol. Macromol. **83**, 61–70 (2016)

M. Casas, C. Ferrero, M.R. Jiménez-Castellanos, Graft tapioca starch copolymers as novel excipients for controlled-release matrix tablets. Carbohydr. Polym. **80**, 71–77 (2010)

A. Femenia, E.S. Sanchez, S. Simal, C. Rossello, Compositional features of polysaccharides from *Aloe Vera* (*Aloe barbadensis* Miller) plant tissues. Carbohydr. Polym. **39**, 109–117 (1999)

G.K. Jani, D.P. Shah, V.C. Jani, Evaluating mucilage from *Aloe barbadensis* Millar as a pharmaceutical excipient for sustained release matrix tablets. Pharm. Technol. **31**, 90–98 (2007)

N.K. Sachan, A. Bhattyacharya, Evaluation of Assam Bora rice starch as a possible natural mucoadhesive polymer in the formulation of microparticulate controlled drug delivery systems. J. Assam. Sci. Soc. **47**, 34–41 (2006)

N.K. Sachan, A. Bhattyacharya, Feasibility of Assam bora rice based matrix microdevices for controlled release of water insoluble drug. Int. J. Pharm. Pharmaceut. Sci. **1**, 96–102 (2009)

B. Singh, V. Sharma, A. Dhiman, M. Devi, Design of *Aloe vera*-alginate gastroretentive drug delivery system to improve the pharmacotherapy. Polym-Plast Technol. Eng. **51**, 1303–1314 (2012)

Conclusion

This volume presents almost all the overviews and recent topics related to various plant polysaccharides-based multiple-unit systems for oral drug delivery applications. Various natural polysaccharides originated from plant parts like gums, starches, and mucilages are abundantly available from the natural resources. On account of the availability of these plant polysaccharides in larger volume from nature, eco-friendly, low extraction expenditure, safety profile, biocompatibility, and biodegradability, these are being explored and exploited to formulate numerous pharmaceutical drug delivery dosage forms. In recent times, a substantial volume of thoughtfulness has been recompensed to formulate numerous oral multiple-unit sustained-release oral drug delivery systems like microparticles, beads, spheroids, etc. The ionotropic-gelation cross-linking as well as covalent cross-linking techniques are being used to formulate the multiple-unit systems made of various plant-derived polysaccharides. Different plant polysaccharides already investigated as biopolymeric excipient polymers to formulate the multiple-unit systems for sustained releasing of drugs are gum Arabica, sterculia gum, locust bean gum, tamarind polysaccharide, okra gum, fenugreek seed mucilage, *Aloe vera* gel, linseed polysaccharide, tapioca starch, potato starch, *etc.*

Printed in the United States
By Bookmasters